T0140370

Studies in Big Data

Volume 4

Series editor

Janusz Kacprzyk, Polish Academy of Sciences, Warsaw, Poland
e-mail: kacprzyk@ibspan.waw.pl

For further volumes:
http://www.springer.com/series/11970

About this Series

The series "Studies in Big Data" (SBD) publishes new developments and advances in the various areas of Big Data-quickly and with a high quality. The intent is to cover the theory, research, development, and applications of Big Data, as embedded in the fields of engineering, computer science, physics, economics and life sciences. The books of the series refer to the analysis and understanding of large, complex, and/or distributed data sets generated from recent digital sources coming from sensors or other physical instruments as well as simulations, crowd sourcing, social networks or other internet transactions, such as emails or video click streams and other. The series contains monographs, lecture notes and edited volumes in Big Data spanning the areas of computational intelligence incl. Neural networks, evolutionary computation, soft computing, fuzzy systems, as well as artificial intelligence, data mining, modern statistics and Operations research, as well as self-organizing systems. Of particular value to both the contributors and the readership are the short publication timeframe and the world-wide distribution, which enable both wide and rapid dissemination of research output.

Weng-Long Chang · Athanasios V. Vasilakos

Molecular Computing

Towards a Novel Computing Architecture
for Complex Problem Solving

 Springer

Weng-Long Chang
Department of Computer Science and
    Information Engineering
National Kaohsiung University
    of Applied Sciences
Kaohsiung
Taiwan
Republic of China

Athanasios V. Vasilakos
Computer Science Department
Kuwait University
Safat
Kuwait

Additional material to this book can be downloaded from http://extras.springer.com.

ISSN 2197-6503   ISSN 2197-6511   (electronic)
ISBN 978-3-319-38098-8   ISBN 978-3-319-05122-2   (eBook)
DOI 10.1007/978-3-319-05122-2
Springer Cham Heidelberg New York Dordrecht London

Printed on acid-free paper

Springer is part of Springer Science+Business Media (www.springer.com)

Preface

In 1946, Electronic Numerical Integrator and Calculator (**ENIAC**) was the first general-purpose, totally electronic computer constructed by John Mauchly and J. Presper Eckert. It contained 18,000 vacuum tubes and was 10 feet high by 100 feet long with 30 tons weight. John von Neumann offered that the program and the data should be stored in memory. Based on von Neumann's ideas, in 1950 at the University of Pennsylvania the first electronic computer that was called EDVAC was constructed. Electronic computers made after 1950 actually follow von Neumann's ideas and vacuum tubes were replaced by the **integrated circuit** which was transistors, wiring, and other components on a single chip.

In 1959, Richard Feynman (Feynman 1961) gave a visionary talk describing the possibility of building computers which were "*sub*-microscopic." In 1994, with Adleman's seminal work (Adleman 1994) the possibility of computing directly with molecules is explored. Since then significant theoretical and experimental results have been reported regularly for the field of molecular computing. Several famous books (Amos 2005; Calude and Paun 2001; Paun et al. 1998) have introduced various paradigms of molecular computing.

A bit which is either 0 or 1 is the smallest unit of data to solve any problem with the input of n bits and 2^n combination states of n bits are all of the possible solutions for that problem. Molecular algorithms are to make use of bit operations to simultaneously operate 2^n combination states and to obtain the required answer(s). The purpose of the book is to teach a reader to how to design various molecular algorithms to construct various kinds of digital arithmetical and logical circuits. Here is an outline of the chapters:

Chapter 1 explains the behaviors of a digital computer and molecular computing. Simultaneously, a concise description to a digital computer and molecular computing based on the von Neumann Architecture is also given.

Chapter 2 introduces a bit pattern which is a uniform representation of data and is also the most efficient solution to how to handle all of the data types. A hexadecimal system with its base that is 16 and an octal system with its base that is eight are concisely illustrated, and an introduction for the conversion between a hexadecimal number or an octal number and a binary number is also given.

Chapter 3 illustrates how eight molecular operations work and use them to develop a parity counter of n bits, the parity generator of error-detection codes on digital communication, and the parity checker of error-detection codes on digital communication.

Chapter 4 describes the conversion between a decimal number and a binary number and develops molecular algorithms to construct the range of the value for an unsigned integer of n bits, a sign-and-magnitude integer of n bits, a one's complement integer of n bits, a two's complement integer of n bits, a floating-point number of n bits in form of single precision format based on Excess_127, and a floating-point number of n bits in the form of double precision format based on Excess_1023.

Chapter 5 develops molecular algorithms to implement a parallel adder of one bit and a parallel adder of n Bits to unsigned integers, and a parallel subtractor of one bit and a parallel subtractor of n Bits for unsigned integers.

Chapter 6 develops molecular algorithms to construct the parallel **NOT** operation of a bit and n bits, the parallel **OR** operation of a bit and n bits, the parallel **AND** operation of a bit and n bits, the parallel **NOR** operation of a bit and n bits, the parallel **NAND** operation of a bit and n bits, the parallel **XOR** operation of a bit and n bits, and the parallel **XNOR** operation of a bit and n bits.

Chapter 7 develops molecular algorithms to complete parallel comparators of a bit and n bits, a parallel left shifter of n bits, a parallel right shifter of n bits, the parallel operation of increase with n bits, the parallel operation of decrease with n bits, and finding the maximum and minimum numbers of one from 2^n combinations of n bits.

The book consists of extensive exercises at the end of each chapter. Solutions of exercises from Chap. 1 through Chap. 7 can easily be completed by readers if they fully understand the contents of each chapter. Instructors, researchers, and students can find the solutions of all the exercises on http://extras.springer.com/.

Power point presentations have been developed for this text as an invaluable tool for learning. Instructors, researchers, and students can find the Power point presentations of all chapters on http://extras.springer.com/.

October 2013   Weng-Long Chang
                                                                              Athanasios V. Vasilakos

References

L.M. Adleman, Molecular computation of solutions to combinatorial problems. Sci. **226**, 1021–1024 (1994)

M. Amos, *Theoretical and Experimental DNA Computation*, (Springer-Verlag, 2005), ISBN: 9783540657736

C.S. Calude, G. Paun, *Computing with Cells and Atoms: An Introduction to Quantum, DNA and Membrane Computing*, (Taylor and Francis, London, 2001), ISBN: 0748408991

R.P. Feynman, in *Miniaturization* (Reinhold Publishing Corporation, New York, 1961), pp. 282–296

G. Paun, G. Rozenberg, A. Salomaa, *DNA Computing: New Computing Paradigms*, (Springer-Verlag, 1998), ISBN: 3540641963

Contents

Chapter 1
Introduction to Digital Computers and Bio-molecular Computer

Today the term "Computer Science" has a very broad meaning. From the viewpoint of computing characteristic, "Computer Science" actually contains a digital computer (Turing 1936, von Neumann 1956), bio-molecular computer (Adleman 1994) and quantum computing (Deutsch 1985). Because the discussion for quantum computing exceeds the scope of this book, thus, we do not introduce "quantum computing". For the purpose of this book, the phrase "bio-molecular computer" describes the in vitro (therefore outside living cell) manipulation of bio-molecules. Those manipulations may be applied to finish various kinds of computations. In this introductory chapter, we try to explain the behaviors of a digital computer and bio-molecular computer.

1.1 The Behaviors of a Digital Computer

If you are not concerned with the internal mechanism of a digital computer, you can simply denote it as a black box. However, you still need to denote the tasks finished by a digital computer for distinguishing it from other types of black boxes. We offer computational model of a digital computer. Figure 1.1 is used to represent computational model of a digital computer.

Fig. 1.1 Computational model of a digital computer

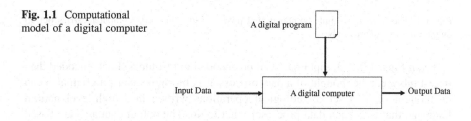

From Fig. 1.1, a digital computer can be thought of as a data processor. A digital program also can be thought of as a set of instructions written in a digital computer language that indicates the data processor what to do with the input data. The output data depend on the combination of two factors: the input data and the digital program. With the same digital program, you can produce different outputs if you change the input. Similarly, with the same input data, you can generate different outputs if you change the digital program.

1.2 The Behaviors of Bio-molecular Computer

For bio-molecular computer, if you are not concerned with the internal mechanism, it can simply be defined as another black box. The black box for bio-molecular computer is shown in Fig. 1.2. In Fig. 1.2, all operations with test tubes have to be carried out by the user. A more advanced model is depicted in Fig. 1.3, where some robotics or electronic computing is used to carry out automatically the majority of the operations with the test tubes without the intervention of the user.

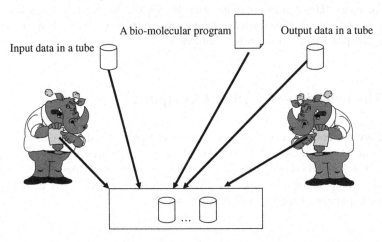

Fig. 1.2 The user in a simple computational model of bio-molecular computer carries out all operations with the test tubes

From Figs. 1.2, 1.3, input data can be encoded in test tubes. Each encoded data in test tubes can be thought of a data processor. A bio-molecular program also can be thought of as a set of biological operations written in a high-level natural language that tells each data processor what to do. The output data also are based on the combination of two factors: the input data and the bio-molecular program. With the same bio-molecular program, you can produce different outputs if you change the input. Similarly, with the same input data, you can generate different outputs if you change the bio-molecular program. Finally, if the input data and the bio-molecular program remain the same, the output should be the same. Let us look at those cases.

In Fig. 1.4, a bio-molecular program is used to find the smallest element for

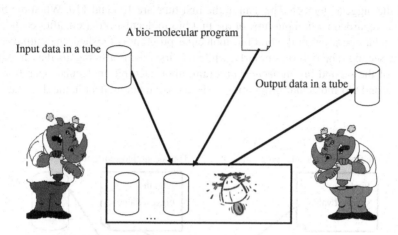

Fig. 1.3 Some robotics or electronic computing in an advanced computational model of bio-molecular computer carries out automatically the majority of the operations with the test tubes without the intervention of the user

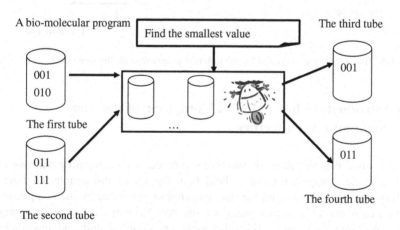

Fig. 1.4 The outputs are generated from the same program with different data

different data in different test tubes. The first tube contains two natural numbers: 001 and 010. The second tube also includes two different natural numbers: 011 and 111. When the first tube is regarded as an input tube of the bio-molecular program, after each bio-molecular operation in the bio-molecular program is performed, the output data in the third tube is 001. Similarly, while the second tube is regarded as an input tube of the bio-molecular program, after each bio-molecular operation in the bio-molecular program is finished, the output data in the fourth tube is 011.

In Fig. 1.5, for the data in the first tube the first bio-molecular program is used to find the smallest element and the second bio-molecular program is applied to find the biggest element. The data in the first tube are 100 and 110. When the first tube is regarded as an input tube of the first bio-molecular program, after each bio-molecular operation in the first bio-molecular program is finished, the output data in the second tube is 100. Similarly, while the first tube is also regarded as an input tube of the second bio-molecular program, after each bio-molecular operation in the second bio-molecular program is performed, the output data in the third tube is 110.

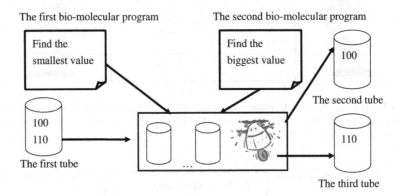

Fig. 1.5   The outputs are generated by the different programs with the same data

1.3 Introduction for a Digital Computer of the von Neumann Architecture

The so-called von Neumann architecture is a model for a computing machine that uses a single storage structure to hold both the set of instructions on how to perform the computation and the data required or generated by the computation. Today, each digital computer based on the von Neumann architecture contains four subsystems: memory, arithmetic logic unit, control unit and input/output devices.

A digital computer system of the von Neumann architecture is shown in Fig. 1.6. From Fig. 1.6, the input subsystem accepts input data and the digital program from outside the digital computer and the output subsystem sends the result of processing to the outside. Memory is the main storage area in the inside of the digital computer system. It is used to store data and digital programs during processing. This implies that both the data and programs should have the same format because they are stored in memory. They are, in fact, stored as binary patterns (a sequence of 0s and 1s) in memory. The arithmetic logic unit is the core of the digital computer system and is applied to perform calculation and logical operations. The control unit is employed to control the operations of the memory, ALU, and the input/output subsystem.

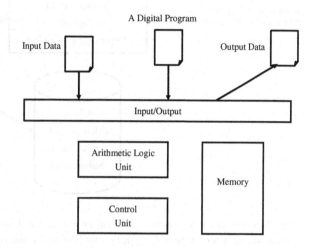

Fig. 1.6 A digital computer system of the von Neumann architecture has four subsystems

A digital program in the von Neumann architecture is made of a finite number of instructions. In the architecture, the control unit fetches one instruction from memory, interprets it, and then excutes it. In other words, the instructions in the digital program are executed one after another. Of course, one instructions may request the control unit to jump to some previous or following one instruction, but this does not mean that the instructions are not executed sequentially.

1.4 The von Neumann Architecture for Bio-molecular Computer

In bio-molecular computer, data also are represented as binary patterns (a sequence of 0s and 1s). Those binary patterns are encoded by sequences of bio-molecules and are stored in a tube. This is to say that a tube is the only storage area in bio-molecular computer and is also the memory and the input/output subsystem of the von Neumann architecture. Bio-molecular programs are made of a set of bio-molecular operations and are used to perform calculation and logical operations. So, bio-molecular programs can be regarded as the arithmetic logic unit of the von Neumann architecture. A robot is used to automatically control the operations of a tube (the memory and the input/output subsystem) and bio-molecular programs (the ALU). This implies that the robot can be regarded as the control unit of the von Neumann architecture.

In Fig. 1.7, bio-molecular computer of the von Neumann architecture is shown. From Fig. 1.7, a robot fetches one bio-molecular operation from a bio-molecular program (the ALU), and then carries out the bio-molecular operation for those data

Fig. 1.7 The bio-molecular computer of the von Neumann architecture

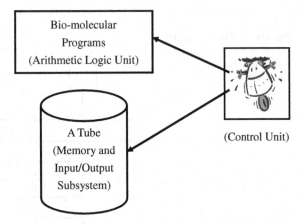

Bio-molecular
Programs
(Arithmetic Logic Unit)

A Tube
(Memory and
Input/Output
Subsystem)

(Control Unit)

stored in the tube (the memory). In other words, the bio-molecular operations are executed one after another. Certainly, one bio-molecular operation perhaps requests the robot to perform some previous or following bio-molecular operations.

1.5 Summary

In this chapter, computational models of "Computer Science" were categorized as a digital computer, bio-molecular computer, quantum computing. The abstract behaviors and the von Neumann architecture of a digital computer were concisely described here. Our attention mainly focused on explaining how the abstract behaviors and the von Neumann architecture of bio-molecular computer worked. An introduction of quantum computing exceeds the scope of this book, so if a general reader is interested in the field of quantum computing, then please refer to a textbook that was written by Imre and Balazs.

1.6 Bibliographical Notes

A popular textbook on foundations of a digital computer is (Forouzan and Mosharraf 2008) in which all areas of the digital computer were covered in breadth. For an excellent introduction to Turing machines and the von Neumann architecture, see Turing (1936) and Neumann (1956). The first article of bio-molecular computer was (Adleman 1994), which solved an instance of the directed Hamiltonian path problem in a graph with seven vertices by means of bio-molecules and biological operations. Practical aspects of constructing a biological computer were introduced in (Adleman 1996; Adleman 1998).

1.7 Exercises

1.1 Explain how a digital computer based on computational model of a black box works.
1.2 Explain how bio-molecular computer based on computational model of another black box works.
1.3 What are respectively a digital program and a bio-molecular program between a digital computer and bio-molecular computer?
1.4 What are subsequently a memory subsystem and a tube between a digital computer and bio-molecular computer?
1.5 What are respectively an input/output subsystem and a tube between a digital computer and bio-molecular computer?
1.6 What are bit 0 and bit 1 to a digital computer and bio-molecular computer?
1.7 Explain how a digital computer based on the von Neumann architecture works.
1.8 Explain how bio-molecular computer based on the von Neumann architecture works.

References

L.M. Adleman, Molecular computation of solutions to combinatorial problems. Science, 226 (1994), pp. 1021–1024
L. M. Adleman, in *On constructing a molecular computer*, eds. by R. Lipton and E. Baum. DNA based computers. DIMACS: series in Discrete Mathematics and Theoretical Computer Science (American Mathematical Society, Providence, 1996), pp. 1–21
L.M. Adleman, Computing with DNA. Sci. Am, **279**(2) (1998), pp. 54–61
D. Deutsch, Quantum theory, the Church-Turing principle and the universal quantum computer. Proc Roy Soc Lond A **400**, 400–497 (1985)
B. Forouzan, F. Mosharraf, in *Foundations of Computer Science*, 2nd edn. (Thomson, London, 2008). ISBN: 978-1-84480-700-0
J. von Neumann, in *Probabilistic Logics and the Synthesis of Reliable Organisms from Unreliable Components* (Princeton University Press, Princeton, 1956), pp. 329–378
A. Turing, On computable numbers, with an application to the entscheidungsproblem. Proc. Lond. Math. Soc. Ser. 2, 42, 230–265 and 43, 544–546 (1936)

Chapter 2
Data Representation on Bio-molecular Computer

The two terms "information" and "data" are alternatively used in this chapter. As discussed in Chap. 1, computing characteristic for bio-molecular computer satisfies the von Neumann architecture. Therefore, bio-molecular computer is an information processing machine. Before we can talk about how to deal with data, you need to fully understand the nature of data. In this chapter, we introduce the different information and how they are stored in tubes in bio-molecular computer.

2.1 Introduction to Data Types

Today information can be represented in different forms of data such as audio, image, video, text, number and so on. Figure 2.1 is used to explain that information is made of five different types of data. From Fig. 2.1, music can be represented in the form of audio data and your voice also can be represented in the form of audio data. Similarly, photos can be regarded as image data and movies can also be regarded as video data. Documents can be generally regarded as text data, and integers that are made of digits are regarded as number data. The term "multimedia" is applied to denote information that includes audio, image, video, text, and number.

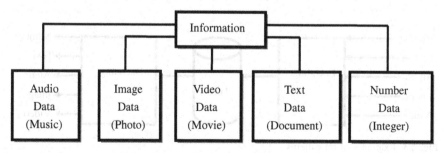

Fig. 2.1 Information is made of five different types of data

W.-L. Chang and A. V. Vasilakos, *Molecular Computing*, Studies in Big Data 4,
DOI: 10.1007/978-3-319-05122-2_2, © Springer International Publishing Switzerland 2014

2.2 Data Representation for Bio-molecular Computer

The interesting question is to how to handle all these data types in Fig. 2.1. Of course, the most efficient solution for the interesting question is to use a uniform representation of data. All data types from outside bio-molecular computer are transformed into this uniform representation when they are stored in tubes in bio-molecular computer, and then transformed back when leaving tubes in bio-molecular computer. This universal format is called a *bit pattern*.

Before further discussion of bit patterns, we must define a bit. A bit (binary digit) is the smallest unit of data that can be stored in tubes in bio-molecular computer; it is either 0 or 1. For a bit in tubes in bio-molecular computer, different sequences of bio-molecules can be used to represent its two states (either 0 or 1). For example, two different sequences of bio-molecules can be regarded as the on state and the off state of a switch in a digital computer. The convention is to represent the on state as 1 and the off state as 0. Therefore, it is very clear that two different sequences of bio-molecules can be applied to represent a bit. In other words, two different sequences of bio-molecules can be employed to store one bit of information. Today, data can be represented different sequences of bio-molecules and also stored in tubes in bio-molecular computer.

A single bit cannot possibly solve the data representation problem. Hence, a bit pattern or a string of bits is used to solve the problem. A bit pattern made of 16 bits is shown in Fig. 2.2. It is a combination of 0s and 1s. This is to say that if a bit pattern made of 16 bits can be stored in a tube in bio-molecular computer, then 32 different sequences of bio-molecules are needed.

1010101001010101

Fig. 2.2 A bit pattern is made of 16 bits

A tube in bio-molecular computer is just used to store the data as bit patterns. It does not know what type of data a stored bit pattern represents. The designer for a bit pattern is responsible for interpreting a bit pattern as number, text or some other

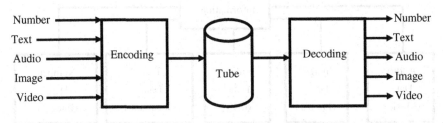

Fig. 2.3 Examples of bit patterns

type of data. In other words, data are encoded when they are stored in a tube and decoded when they are presented to the designer (Fig. 2.3). By tradition, a bit pattern of length 8 is called a byte. This term is used to measure the size of data stored in a tube in bio-molecular computer. For example, generally speaking, a tube in bio-molecular computer that can be applied to store 10^{15} bits of information is said to have 1.25×10^{14} bytes of information.

2.3 Hexadecimal Notation

The bit pattern is designed to represent data when they are stored in tubes in bio-molecular computer. However, to manipulate bit patterns is found to be difficulty for people. Using a long stream of 0s and 1s is tedious and prone to error. Hexadecimal notation is applied to improve this situation. Hexadecimal notation is based on 16 (hexadec is Greek for 16). This implies that 16 symbols (hexadecimal digits): 0, 1, 2, 3, 4, 5, 6, 7, 8, 9, A, B, C, D, E, and F. Each hexadecimal digit can be represented by four bits, and four bits also can be represented by a hexadecimal digit. The relationship between a bit pattern and a hexadecimal digit is shown in Table 2.1.

	Bit pattern	Hexadecimal digit	Bit pattern	Hexadecimal digit
Table 2.1 The corresponding table has the relation among hexadecimal digits and binary digits	0000	0	1000	8
	0001	1	1001	9
	0010	2	1010	A
	0011	3	1011	B
	0100	4	1100	C
	0101	5	1101	D
	0110	6	1110	E
	0111	7	1111	F

Converting from a bit pattern to hexadecimal is done by organizing the pattern into groups of four and finding the hexadecimal value for each group of 4 bits. For hexadecimal to bit pattern conversion, convert each hexadecimal digit to its 4-bit equivalent (Fig. 2.4). Generally speaking, hexadecimal notation is written in two formats. In the first format, a lowercase (or uppercase) x is added before the digits to show that the representation is in hexadecimal. For example, xCFD8 is applied to represent a hexadecimal value in this convention. In another format, the base of the number (16) is indicated as the subscript after the notation. For example, $(CFD8)_{16}$ shows the same value in the second convention. In this book, we use both conventions.

Fig. 2.4 Binary to
hexadecimal and
hexadecimal to binary
transformation

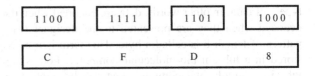

2.4 Octal Notation

Another notation used to group bit patterns together is octal notation. Octal
notation is based on 8 (*oct* is Greek for 8). This implies that there are eight
symbols (octal digits): 0, 1, 2, 3, 4, 5, 6, and 7. An octal digit can represent three
bits, and three bits can be represented by an octal digit. The relationship between a
bit pattern and an octal digit is shown in Table 2.2. From Table 2.2, the first bit
pattern 000 corresponds to the first octal digit 0, the second bit pattern 001 corresponds to the second octal digit 1, and so on with that the last bit pattern 111
corresponds to the last octal digit 7.

Table 2.2 The
corresponding table has the
relation among octal digits
and binary digits

Bit pattern	Octal digit	Bit pattern	Octal digit
000	0	100	4
001	1	101	5
010	2	101	6
011	3	111	7

Converting from a bit pattern to octal is performed through organizing the
pattern into groups of three and finding the octal value for each group of three bits.
For octal to bit pattern conversion, convert each octal digit to its 3-bit equivalent
(Fig. 2.5). Generally speaking, octal notation is written in two formats. In the first
format, a 0 (zero) is added before the digits to show that the representation is in
octal. For example, the value, 04756, is applied to represent an octal value in this
convention. In another format, the base of the number (8) is indicated as the
subscript after the notation. For example, $(4756)_8$ shows the same value in the
second convention. In this book, we use both conventions.

Fig. 2.5 Binary to octal and
octal to binary transformation

1 0 0	1 1 1	1 0 1	1 1 0

4	7	5	6

2.5 Summary

In this chapter, an introduction to data representation of bio-molecular computer was provided. We described information that was made of different types of data and used a bit pattern as a uniform representation of data. We then introduced a bit that is the smallest unit of data that can be stored in tubes in bio-molecular computer, and used different sequences of bio-molecules to encode its two states (either 0 or 1). We also introduced a designer to a bit pattern that was responsible for interpreting a bit pattern. We then described a hexadecimal system in which its base is 16 and we used sixteen symbols to represent numbers. Simultaneously, we also introduced conversion from a binary system to a hexadecimal system and from a hexadecimal system to a binary system. We then described an octal system in which its base is 8 and we used eight symbols to represent numbers. Similarly, we also introduced conversion from a binary system to an octal system and from an octal system to a binary system.

2.6 Bibliographical Notes

In this chapter for more details about data types in a digital computer, the recommended books are Forouzan and Mosharraf (2008); Koren (2001); Marques and Silva (2012); Miano (1999). For a more detailed introduction to data representation in bio-molecular, the recommended books are Amos (2005); Ehrenfeucht et al. (2004); Paun et al. (1998). For more details about the subjects of a number system discussed in a digital computer, the recommended books are Forouzan and Mosharraf (2008); Mano (1979); Reed (2008); Shiva (2008).

2.7 Exercises

2.1. For a digital computer and bio-molecular computer, a bit is the smallest unit of data in which its vale either 1 or 0. Answer the following questions about how to encoding a bit:

   a. How are to a bit its values 0 and 1 encoded in a digital computer?
   b. How are to a bit its values 0 and 1 encoded in bio-molecular computer?

2.2. A bit pattern is a uniform representation of information. Answer the following questions about how to encoding a bit pattern:

   a. How is a bit pattern encoded in a digital computer?
   b. How is a bit pattern encoded in bio-molecular computer?

2.3 Write a program of a digital computer to convert a *hexadecimal* number to its corresponding *binary* number.

2.4 Write a program of a digital computer to convert a *binary* number to its corresponding *hexadecimal* number.

2.5 Write a program of a digital computer to convert an *octal* number to its corresponding *binary* number.

2.6 Write a program of a digital computer to convert a *binary* number to its corresponding *octal* number.

References

A. Ehrenfeucht, T.H.I. Petre, D.M. Prescott, G. Rozenberg, *Computation in Living Cells: Gene Assembly in Ciliates*. (Springer, Hidelberg, 2004). ISBN: 3540407952

B. Forouzan, F. Mosharraf, *Foundations of Computer Science*, 2nd edn. (Thomson, London, 2008). ISBN: 978-1-84480-700-0)

D. Reed, *A Balanced Introduction to Computer Science*. (Pearson Prentice Hall, New Jersy, 2008). ISBN: 9780136017226

G. Paun, G. Rozenberg, A. Salomaa, *DNA Computing: New Computing Paradigms*. (Springer, Hidelberg, 1998). ISBN: 3540641963

I. Koren, *Computer Arithmetic Algorithms*. (A k Peters, Natick, 2001). ISBN: 1568811608

J. Miano, *Compressed Image File Formats: JPEG, PNG, GIF, XBM, BMP*. (Addison Wesley, Boston, 1999). ISBN: 0201604434

M. Marques, D. Silva, *Multimedia Communications and Networking*. (CRC Press, New York, 2012). ISBN: 978-1439874844

M. Amos, *Theoretical and Experimental DNA Computation*. (Springer, Hidelberg, 2005). ISBN: 9783540657736

M.M. Mano, *Digital Logic and Computer Design*. (Prentice-Hall, New Jersy, 1979). ISBN: 0-13-214510-3

S.G. Shiva, *Computer Organization, Design, and Architecture*. (CRC Press, Boca Raton, 2008). ISBN: 9780849304163

Chapter 3
Introduction for Bio-molecular Operations on Bio-molecular Computer

In this chapter, we first introduce how eight bio-molecular operations are used to perform representation of bit patterns for data stored in tubes in bio-molecular computer. Then, we describe how eight bio-molecular operations are applied to deal with various problems.

3.1 Introduction to Bio-molecular Operations

A set P is equal to $\{x_n \ldots x_1|$ each x_k is a binary value for $1 \leq k \leq n$, where $n \geq 0\}$. This means that if n is not equal to zero, then the set P is not an empty set. Otherwise, it is an empty set. A tube is a storage device for bio-molecular computer. Data represented by bit patterns are stored in the tube. Therefore, a set P can be regarded as a tube T and an element in the set P can also be regarded as one data stored in the tube T. This is to say that for the kth bit of a data stored in a tube T, x_k, two *distinct* sequences of bio-molecules are designed to represent its two states (either 0 0r 1). One represents the value "0" for x_k and the other represents the value "1" for x_k. For convenience, x_k^1 is applied to denote the value of x_k to be 1 and x_k^0 is used to define the value of x_k to be 0.

The corresponding bio-molecular programs perform computational tasks for any data in a tube. We define that any bio-molecular program must be made of a combination of only these three constructs: sequence, decision (selection), and repetition (Fig. 3.1). The first construct in Fig. 3.1 is called the sequence construct. Any bio-molecular program eventually is a sequence of bio-molecular operations, which can be a bio-molecular operation or either of the other two constructs. Some questions can be solved with testing some different conditions. Therefore, if the result of a tested condition is true, then a bio-molecular program follows a sequence of bio-molecular operations. Otherwise, a bio-molecular program follows a different sequence of bio-molecular operations. This is called the decision (selection) construct and it is shown in Fig. 3.1. For solving some other problems, the same sequence of bio-molecular operations must be repeated. The repetition construct shown in Fig. 3.1 is applied to handle this.

W.-L. Chang and A. V. Vasilakos, *Molecular Computing*, Studies in Big Data 4, DOI: 10.1007/978-3-319-05122-2_3, © Springer International Publishing Switzerland 2014

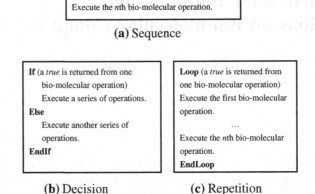

(a) Sequence

(b) Decision   (c) Repetition

Fig. 3.1 Three constructs of a bio-molecular program

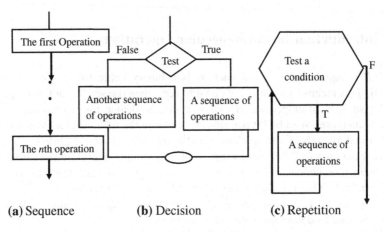

(a) Sequence   (b) Decision   (c) Repetition

Fig. 3.2 Flowcharts for three constructs of a bio-molecular program

A flowchart is a pictorial representation of a bio-molecular program. It hides all of the details of a bio-molecular program in an attempt to give the big picture; it shows how the bio-molecular program flows from beginning to end. The three constructs in Fig. 3.1 are represented in flowcharts (Fig. 3.2).

3.2 Introduction to the Append Operation

Eight bio-molecular operations are used to deal with each data in a tube *T*. Each bio-molecular program is made of eight bio-molecular operations. The first operation is the *append* operation. **Definition 3-1** is applied to describe how the *append* operation deals with data in a tube.

Definition 3-1: Given a tube T and a binary digit x_j, the operation, "*Append*", will append x_j onto the end of every data stored in the tube T. The formal representation for the operation is written as "Append(T, x_j)".

Any tube is initialized to be an empty tube, so there is no data stored in it. If we want the bit pattern, $x_n \ldots x_1$, to be stored in a tube T, then the *append* operation can be applied to perform the task. Therefore, in light of three constructs in Fig. 3.1, the following simple bio-molecular program can be applied to construct the bit pattern, $x_n \ldots x_1$, to be stored in a tube T.

Algorithm 3.1: ConstructBitPattern(T, n)

(1) **For** $k = n$ **downto** 1

      (1a) Append(T, x_k).

   EndFor

EndAlgorithm

A tube T is an empty tube and is regarded as the input tube for the algorithm, **ConstructBitPattern(T, n)**. For the second parameter n in **Algorithm 3.1**, it is used to denote the number of bit for a bit pattern. Step (1) in **Algorithm 3.1** is applied to represent the repetition construct in Fig. 3.1 and to denote the number of execution for Step (1a). Step (1a) in **Algorithm 3.1** is used to represent the sequence construct in Fig. 3.1 and is made of the *append* operation denoted in Definition **3-1**. After the first execution of Step (1a) is performed, the bit x_n is stored in the tube T. Next, after the second execution of Step (1a) is implemented, the two bits $x_n x_{n-1}$ is stored in the tube T. After repeating to execute n times for Step (1a), the result for the tube T is shown in Table 3.1.

Table 3.1 The result for the tube T is generated by Algorithm 3.1	Tube	The result is generated by **Algorithm 3.1**
	T	$\{x_n x_{n-1} \ldots x_1\}$

Lemma 3-1: *The algorithm,* **ConstructBitPattern(T, n)**, *can be used to construct the bit pattern,* $x_n \ldots x_1$, *to be stored in a tube T.*

Proof The algorithm, **ConstructBitPattern(T, n)**, is implemented by means of the *append* operation. Each execution of Step (1a) is used to append the value "1" for x_k or the value "0" for x_k onto the end of $x_n \ldots x_{k+1}$ in tube T. This implies that the kth bit in the bit pattern, $x_n x_{n-1} \ldots x_1$, is stored in the tube T. Therefore, it is inferred that the algorithm, **ConstructBitPattern(T, n)**, can be used to construct the bit pattern, $x_n \ldots x_1$, to be stored in a tube T. ∎

3.3 Introduction to the Amplify Operation

The second bio-molecular operation is the *amplify* operation. It is used to perform the copy of data stored in any a tube. **Definition 3-2** is applied to describe how the amplify operation manipulates data stored in any a tube.

Definition 3-2: Given a tube T, the operation "Amplify(T, T_1, T_2)" will produce two new tubes T_1 and T_2 so that T_1 and T_2 are totally a copy of T (T_1 and T_2 are now identical) and T becomes an empty tube.

If we want to generate the same two bit patterns, $x_n \ldots x_1$, then the following bio-molecular program can be used to perform our requirement. Three tubes T, T_1, and T_2 are empty tubes and are regarded as input tubes of **Algorithm 3.2**. For n, it is denoted as the number of bits for $x_n \ldots x_1$ and is regarded as the fourth parameter of **Algorithm 3.2**.

> **Algorithm 3.2: CopyBitPattern(T, T_1, T_2, n)**
>
> (1) **ConstructBitPattern(T, n)**.
>
> (2) Amplify(T, T_1, T_2).
>
> **EndAlgorithm**

Step (1) in **Algorithm 3.2** is employed to call **Algorithm 3.1** for producing a bit pattern $x_n \ldots x_1$ stored in the tube T. Then, on the execution of Step (2), the *amplify* operation is used to generate two new tubes T_1 and T_2 containing the same bit pattern $x_n \ldots x_1$ and the tube T becomes an empty tube. The result is shown in Table 3.2 after each operation in **Algorithm 3.2** is performed.

Table 3.2 Algorithm 3.2 generates the result	Tube	The result is produced by **Algorithm 3.2**
	T	$\varnothing$
	T_1	$\{x_n\, x_{n-1} \ldots x_1\}$
	T_2	$\{x_n\, x_{n-1} \ldots x_1\}$

Lemma 3-2: *The algorithm,* **CopyBitPattern**(T, T_1, T_2, n), *can be applied to generate the same two bit patterns,* $x_n \ldots x_1$.

Proof Refer to **Lemma 3-1**.

3.4 Introduction to the Merge Operation

The third bio-molecular operation is the *merge* operation. It is employed to perform the merge of data stored in any n tubes. **Definition 3-3** is used to describe how the *merge* operation pours data stored in any n tubes into one tube.

Definition 3-3: Given n tubes $T_1 \ldots T_n$, the *merge* operation is to pour data stored in any n tubes into one tube, without any change in the individual data. The formal representation for the *merge* operation is written as "$\cup(T_1, \ldots, T_n)$", where $\cup(T_1, \ldots, T_n) = T_1 \cup \ldots \cup T_n$.

The value of each bit in a bit pattern $x_n \ldots x_1$ is either 0 or 1. Because each bit has two states, n bits can be used to generate 2^n combinational states. The following bio-molecular program can be employed to produce 2^n combinational states. A tube T_0 is an empty tube and is regarded as the input tube of **Algorithm 3.3**. For the second parameter n in **Algorithm 3.3**, it is applied to represent the number of bits.

Algorithm 3.3: CombinationalStates(T_0, n)

(0a) Append(T_1, x_n^1).

(0b) Append(T_2, x_n^0).

(0c) $T_0 = \cup(T_1, T_2)$.

(1) **For** $k = n - 1$ **downto** 1

    (1a) Amplify(T_0, T_1, T_2).

    (1b) Append(T_1, x_k^1).

    (1c) Append(T_2, x_k^0).

    (1d) $T_0 = \cup(T_1, T_2)$.

  EndFor

EndAlgorithm

Consider that eight states for a bit pattern, $x_3\, x_2\, x_1$, are, respectively, 000, 001, 010, 011, 100, 101, 110 and 111. Tube T_0 is an empty tube and is regarded as an input tube of **Algorithm 3.3**. Because the value for n is three, when the execution of Step (0a) and the execution of Step (0b) are finished, tube $T_1 = \{x_3^1\}$ and tube $T_2 = \{x_3^0\}$. Then, on the execution of Step (0c), it uses the *merge* operation to pour tubes T_1 and T_2 into tube T_0. This implies that $T_0 = \{x_3^1, x_3^0\}$, tube $T_1 = \varnothing$, and tube $= \varnothing$. Since Step (1) is the only loop and the value for n is three, Steps (1a) through (1d) will be run two times.

After the first execution of Step (1a) is finished, tube $T_0 = \varnothing$, tube $T_1 = \{x_3^1, x_3^0\}$ and tube $T_2 = \{x_3^1, x_3^0\}$. Next, after the first execution for Step (1b) and Step (1c) is performed, tube $T_1 = \{x_3^1 x_2^1, x_3^0 x_2^1\}$ and tube $T_2 = \{x_3^1 x_2^0, x_3^0 x_2^0\}$. After the first execution of Step (1d) is implemented, tube $T_0 = \{x_3^1 x_2^1, x_3^0 x_2^1, x_3^1 x_2^0, x_3^0 x_2^0\}$, tube $T_1 = \varnothing$ and tube $T_2 = \varnothing$.

Then, after the second execution of Step (1a) is finished, tube $T_0 = \varnothing$, tube $T_1 = \{x_3^1 x_2^1, x_3^0 x_2^1, x_3^1 x_2^0, x_3^0 x_2^0\}$ and tube $T_2 = \{x_3^1 x_2^1, x_3^0 x_2^1, x_3^1 x_2^0, x_3^0 x_2^0\}$. After the rest of operations are performed, the result is shown in Table 3.3. **Lemma 3-3** is applied to demonstrate correction of **Algorithm 3.3**.

Table 3.3 The result for tube T_0 is generated by **Algorithm 3.3**

Tube	The result is generated by **Algorithm 3.3**
T_0	$\{x_3^1 x_2^1 x_1^1,\ x_3^1 x_2^1 x_1^0,\ x_3^1 x_2^0 x_1^1,\ x_3^1 x_2^0 x_1^0,$ $x_3^0 x_2^1 x_1^1,\ x_3^0 x_2^1 x_1^0,\ x_3^0 x_2^0 x_1^1,\ x_3^0 x_2^0 x_1^0\}$
T_1	$\varnothing$
T_2	$\varnothing$

Lemma 3-3: **Algorithm 3.3** *can be applied to construct* 2^n*combinational states of n bits.*

Proof **Algorithm 3.3** is implemented by means of the *amplify, append* and *merge* operations. Each execution of Step (0a) and each execution of Step (0b), respectively, append the value "1" for x_n as the first bit of every data stored in tube T_1 and the value "0" for x_n as the first bit of every data stored in tube T_2. Next, each execution of Step (0c) is to pour tubes T_1 and T_2 into tube T_0. This implies that tube T_0 contains all of the data that have $x_n = 1$ and $x_n = 0$ and tubes T_1 and T_2 become empty tubes.

Each execution of Step (1a) is used to amplify tube T_0 and to generate two new tubes, T_1 and T_2, which are copies of T_0. Tube T_0 then becomes empty. Then, each execution of Step (1b) appends the value "1" for x_k onto the end of $x_n \ldots x_{k+1}$ in every data stored in tube T_1. Similarly, each execution of Step (1c) also appends the value "0" for x_k onto the end of $x_n \ldots x_{k+1}$ in every data stored in tube T_2. Next, each execution of Step (1d) pours tubes T_1 and T_2 into tube T_0. This indicates that data stored in tube T_0 include $x_k = 1$ and $x_k = 0$. After repeating Steps (1a) through (1d), tube T_0 consists of 2^n combinational states of n bits. ∎

3.5 Introduction to the Rest of Bio-molecular Operations

The rest of bio-molecular operations are, subsequently, the *extract* operation, the *detect* operation, the *discard* operation, the *append-head* operation and the *read* operation. **Definitions 3-4** through **3-8** are employed to describe how the rest of bio-molecular operations deal with data stored in any a tube.

Definition 3-4: Given a tube T and a binary digit x_k, the *extract* operation will produce two tubes $+(T, x_k)$ and $-(T, x_k)$, where $+(T, x_k)$ is all of the data in T which contain x_k and $-(T, x_k)$ is all of the data in T which do not contain x_k.

Definition 3-5: Given a tube T, the *detect* operation is used to check whether any a data is included in T or not. If at least one data is included in T we have "yes", and if no data is included in T we have "no". The formal representation for the operation is written as "Detect(T)".

Definition 3-6: Given a tube T, the *discard* operation will discard T. The formal representation for the operation is written as "Discard(T)" or "$T = \varnothing$".

Definition 3-7: Given a tube T and a binary digit x_j, the operation, "*Append-head*", will append x_j onto the head of every data stored in the tube T. The formal representation for the operation is written as "Append-head(T, x_j)".

Definition 3-8: Given a tube T, the *read* operation is used to describe any a data, which is contained in T. Even if T contains many different data, the operation can give an explicit description of exactly one of them. The formal representation for the operation is written as "read(T)".

A one-bit parity counter is to count whether the number of 1's for two input bits is odd or even or not. It includes two inputs and one output. The first input bit is used to represent the current bit to be checked whether the number of 1's is odd or even or not. The second input is used to represent the parity from the previous lower significant position. The first output gives the current value of the parity. The truth table of a one-bit parity counter is shown in Table 3.4.

Table 3.4 The truth table of a one-bit parity counter is shown

The first input	The second input	The first output
0	0	0
0	1	1
1	0	1
1	1	0

One one-bit binary number y_g is used to represent the first input of a one-bit parity counter for $1 \le g \le n$, and two one-bit binary numbers, z_g and z_{g-1}, are applied to represent the first output and the second input of a one-bit parity counter, respectively. For convenience, z_g^1, z_g^0, z_{g-1}^1, z_{g-1}^0, y_g^1 and y_g^0, subsequently, contain the value "1" of z_g, the value "0" of z_g, the value "1" of z_{g-1}, the value "0" of z_{g-1}, the value "1" of y_g, and the value "0" of y_g. The following bio-molecular program can be used to construct a parity counter.

Algorithm 3.4: OneBitParityCounter(T_0, g)

(1) $T_1 = +(T_0, y_g^1)$ and $T_2 = -(T_0, y_g^1)$.

(2) $T_3 = +(T_1, z_{g-1}^1)$ and $T_4 = -(T_1, z_{g-1}^1)$.

(3) $T_5 = +(T_2, z_{g-1}^1)$ and $T_6 = -(T_2, z_{g-1}^1)$.

(4) If (Detect(T_3) = "yes") then

    (4a) Append-head(T_3, z_g^0).

  EndIf

(5) If (Detect(T_4) = "yes") then

    (5a) Append-head(T_4, z_g^1).

  EndIf

(6) If (Detect(T_5) = "yes") then

    (6a) Append-head(T_5, z_g^1).

  EndIf

(7) If (Detect(T_6) = "yes") then

    (7a) Append-head(T_6, z_g^0).

  EndIf

(8) $T_0 = \cup(T_3, T_4, T_5, T_6)$.

EndAlgorithm

Lemma 3-4: Algorithm 3.4 *can be used to perform the function of a one-bit parity counter.*

Proof **Algorithm 3.4** is implemented by means of the *extract, append-head, detect* and *merge* operations. Each execution for Steps (1) through (3) employs the *extract* operations to form some different tubes. This implies that tube T_3 includes all of data that have $y_g = 1$ and $z_{g-1} = 1$, tube T_4 contains all of data that have $y_g = 1$ and $z_{g-1} = 0$, tube T_5 consists of all of data that have $y_g = 0$ and $z_{g-1} = 1$, tube T_6 includes all of data that have $y_g = 0$ and $z_{g-1} = 0$, tube $T_0 = \varnothing$, tube $T_1 = \varnothing$, and tube $T_2 = \varnothing$.

Next, Steps (4), (5), (6) and (7) are, respectively, used to check whether contains any a data for tubes T_3, T_4, T_5, and T_6 or not. If any a "yes" is returned for those steps, then the corresponding *append-head* operations will be run. On each execution of Steps (4a), (5a), (6a) and (7a), the *append-head* operations are employed to respectively pour four different outputs of a one-bit parity counter in Table 3.4 into tubes T_3 through T_6. Finally, each execution of Step (8) applies the *merge* operation to pour tubes T_3 through T_6 into tube T_0. Tube T_0 contains all of the data finishing the function of a one-bit parity counter. ∎

3.6 The Construction of a Parity Counter of N Bits

The one-bit parity counter introduced in Sect. 3.5 is to count whether the number of 1's for two input bits is odd or even or not. A parity counter of n bits can be used to count whether the number of 1's for 2^n combinational states is odd or even by means of n times of this one-bit parity counter. The following algorithm is proposed to finish the function of a parity counter of n bits. A tube T_0 is an empty tube and is regarded as the input tube of **Algorithm 3.5**. For the second parameter n in **Algorithm 3.5**, it is applied to represent the number of bits.

Algorithm 3.5: ParityCounter(T_0, n)

(0a) Append-head(T_1, y_1^1).

(0b) Append-head(T_2, y_1^0).

(0c) $T_0 = \cup(T_1, T_2)$.

(1) **For** $g = 2$ **to** n

    (1a) Amplify(T_0, T_1, T_2).

    (1b) Append-head(T_1, y_g^1).

    (1c) Append-head(T_2, y_g^0).

    (1d) $T_0 = \cup(T_1, T_2)$.

EndFor

(2) Append-head(T_0, z_0^0).

(3) **For** $g = 1$ **to** n

    (3a) **OneBitParityCounter(T_0, g)**

EndFor

EndAlgorithm

Lemma 3-5: Algorithm 3.5 *can be applied to finish the function of a parity counter of n bits.*

Proof **Algorithm 3.5** is implemented by means of the *extract, append-head, detect, amplify,* and *merge* operations. Each execution of Step (0a) and each execution of Step (0b), respectively, append the value "1" for y_1 as the first bit of every data stored in tube T_1 and the value "0" for y_1 as the first bit of every data stored in tube T_2. Next, each execution of Step (0c) is to pour tubes T_1 and T_2 into tube T_0. This implies that tube T_0 contains all of the data that have $y_1 = 1$ and $y_1 = 0$ and tubes T_1 and T_2 become empty tubes.

Step (1) is the first loop and is mainly applied to generate 2^n combinational states. Each execution of Step (1a) is used to amplify tube T_0 and to generate two new tubes, T_1 and T_2, which are copies of T_0. Tube T_0 then becomes empty. Then, each execution of Step (1b) appends the value "1" for y_g onto the head of $y_{g-1} \ldots y_1$ in every data stored in tube T_1. Similarly, each execution of Step (1c) also appends the value "0" for y_g onto the head of $y_{g-1} \ldots y_1$ in every data stored in tube T_2. Next, each execution of Step (1d) pours tubes T_1 and T_2 into tube T_0. This indicates that data stored in tube T_0 include $y_g = 1$ and $y_g = 0$. After repeating Steps (1a) through (1d), tube T_0 consists of 2^n combinational states of n bits.

Because a one-bit parity counter deals with the parity of y_1, the second input must be zero. Therefore, each execution of Step (2) uses the *append-head* operation to append the value "0" of z_0 into the head of each bit pattern in 2^n combinational states. Step (3) is the second loop and is mainly used to finish the function of a parity counter of n bits. On each execution of Step (3a), it calls **Algorithm 3.4** to perform the function of a one-bit parity counter. Repeat to execute Step (3a) until the nth bit, y_n, in each bit pattern is processed. This is to say that tube T_0 contains 2^n combinational states in which each combinational state performs the function of a parity counter of n bits. ∎

3.7 The Power for a Parity Counter of *N* Bits

Consider that four states for a bit pattern, $y_2 y_1$, are, respectively, 00, 01, 10 and 11. Tube T_0 is an empty tube and is regarded as an input tube of **Algorithm 3.5**. After the first execution of Step (0a) and the first execution of Step (0b) are performed, tube $T_1 = \{y_1^1\}$ and tube $T_2 = \{y_1^0\}$. Next, the first execution of Step (0c) is finished, tube $T_0 = \{y_1^1, y_1^0\}$, tube $T_1 = \varnothing$ and tube $T_2 = \varnothing$.

Because the value for n is two, Steps (1a) through (1d) will be run one time. After the first execution of Step (1a) is implemented, tube $T_0 = \varnothing$, tube $T_1 = \{y_1^1, y_1^0\}$ and tube $T_2 = \{y_1^1, y_1^0\}$. Next, after the first execution for Step (1b) and Step (1c) is performed, tube $T_1 = \{y_2^1 y_1^1, y_2^1 y_1^0\}$ and tube $T_2 = \{y_2^0 y_1^1, y_2^0 y_1^0\}$. After the first execution of Step (1d) is implemented, tube $T_0 = \{y_2^1 y_1^1, y_2^1 y_1^0, y_2^0 y_1^1, y_2^0 y_1^0\}$, tube $T_1 = \varnothing$ and tube $T_2 = \varnothing$. Then, after each execution of Step (2) is performed, tube $T_0 = \{z_0^0 y_2^1 y_1^1, z_0^0 y_2^1 y_1^0, z_0^0 y_2^0 y_1^1, z_0^0 y_2^0 y_1^0\}$. Because the value of the upper bound in Step (3) is two, **Algorithm 3.4**, **OneBitParityCounter**(T_0, g), in Step (3a) will be invoked two times.

When the first time for **Algorithm 3.4** in Sect. 3.5 is invoked by **Algorithm 3.5** in Sect. 3.6, tube $T_0 = \{z_0^0 y_2^1 y_1^1, z_0^0 y_2^0 y_1^0, z_0^0 y_2^0 y_1^1, z_0^0 y_2^0 y_1^0\}$ and it is regarded as an input tube to **Algorithm 3.4**. The value for g is one and it is regarded as the second parameter in **Algorithm 3.4**. After the first execution of Step (1) in **Algorithm 3.4** is implemented, tube $T_1 = \{z_0^0 y_2^1 y_1^1, z_0^0 y_2^0 y_1^1\}$, tube $T_2 = \{z_0^0 y_2^1 y_1^0, z_0^0 y_2^0 y_1^0\}$, and tube $T_0 = \varnothing$. Next, after the first execution for Steps (2) and (3) is performed, tube $T_3 = \varnothing$, tube $T_5 = \varnothing$, tube $T_4 = \{z_0^0 y_2^1 y_1^1, z_0^0 y_2^0 y_1^1\}$, and tube $T_6 = \{z_0^0 y_2^1 y_1^0, z_0^0 y_2^0 y_1^0\}$. After a "no" from the first execution of Step (4) is returned, so the first execution of Step (4a) is not run. Then, after a "yes" from the first execution of Step (5) is returned, so the first execution of Step (5a) is implemented and tube $T_4 = \{z_1^1 z_0^0 y_2^1 y_1^1, z_1^1 z_0^0 y_2^0 y_1^1\}$. After a "no" from the first execution of Step (6) is returned, so the first execution of Step (6a) is not run. Then, after a "yes" from the first execution of Step (7) is returned, so the first execution of Step (7a) is implemented and tube $T_6 = \{z_1^0 z_0^0 y_2^1 y_1^0, z_1^0 z_0^0 y_2^0 y_1^0\}$. Finally, after the first execution of Step (8) is finished, the result is shown in Table 3.5 and the first execution of **Algorithm 3.4** is terminated. Then, when the second execution for Step (3a) in **Algorithm 3.5** is implemented, the final result is shown in Table 3.6 and **Algorithm 3.5** is terminated.

Table 3.5 **Algorithm 3.4** generates the result	Tube	**Algorithm 3.4** generates the result
	T_0	$\{z_1^1 z_0^0 y_2^1 y_1^1, z_1^1 z_0^0 y_2^0 y_1^1, z_1^1 z_0^0 y_2^1 y_1^0, z_1^0 z_0^0 y_2^0 y_1^0\}$

Table 3.6 **Algorithm 3.5** generates the result	Tube	**Algorithm 3.5** generates the result
	T_0	$\{z_2^0 z_1^1 z_0^0 y_2^1 y_1^1, z_2^0 z_1^1 z_0^0 y_2^0 y_1^1, z_2^1 z_1^0 z_0^0 y_2^1 y_1^0, z_2^0 z_1^0 z_0^0 y_2^0 y_1^0\}$

3.8 Introduction for the Parity Generator of Error-Detection Codes on Digital Communication

On digital computer systems, binary information may be transmitted through some form of communication medium such as radio waves or wires. A physical communication medium changes bit values either from 1 to 0 or from 0 to 1 if it is disturbed from any external noise. An error-detection code can be applied to detect errors during transmission. The detected error cannot be corrected, but its present is pointed out.

For digital computer systems, during transfer of information from one location to another location, in sending end a "parity-generation" is used to generate the corresponding parity bit for it and in receiving end a "parity-checker" is applied to check the proper parity adopted. An error is detected if the checked parity does not correspond to the adopted one. The parity method can be employed to detect the presence of one, three, or any odd combination of errors. However, even combination of errors is undetectable.

From **Algorithm 3.5**, it is clearly determined whether the number of 1's for 2^n combinational states is even or odd. We use the amplify operation, "Amplify(T_0, T_0^S, T_0^R)", to generate two new tubes T_0^S and T_0^R so that T_0^S and T_0^R are totally a copy of T_0, where tube T_0 is generated from **Algorithm 3.5**. Tubes T_0^S and T_0^R are, respectively, put in the sending end and in receiving end. **Algorithm 3.6** can be applied to replace logic circuits of a "parity-generator" in sending end. Tube T_0^S is regarded as an input tube of **Algorithm 3.6**. The second parameter, n, in **Algorithm 3.6** is the number of bits for transmitted messages. In **Algorithm 3.6**, the third parameter, tube T_{Input}^S, is applied to store any message transmitted. Similarly, in **Algorithm 3.6**, the fourth parameter, tube T_{Output}^S, is used to store those transmitted messages, in which each transmitted message contains the corresponding parity bit.

Algorithm 3.6: ParityGeneration(T_0^S, n, T_{Input}^S, T_{Output}^S)

(1) **For** $g = 1$ **to** n

    (1a) $T_1^{ON} = +(T_0^S, y_g^1)$ and $T_1^{OFF} = -(T_0^S, y_g^1)$.

    (1b) $T_2^{ON} = +(T_{Input}^S, y_g^1)$ and $T_2^{OFF} = (T_{Input}^S, y_g^1)$.

    (1c) **If** (Detect(T_2^{ON}) = "yes") **then**

        (1d) $T_0^S = \cup(T_0^S, T_1^{ON})$ and $T_3 = \cup(T_3, T_1^{OFF})$.

    Else

        (1e) $T_0^S = \cup(T_0^S, T_1^{OFF})$ and $T_3 = \cup(T_3, T_1^{ON})$.

    EndIf

    (1f) $T_{Input}^S = (T_2^{ON}, T_2^{OFF})$.

EndFor

(2) $T_4^{ON} = +(T_0^S, z_n^1)$ and $T_4^{OFF} = -(T_0^S, z_n^1)$.

(3) $T_{Output}^S = \cup(T_{Output}^S, T_{Input}^S)$.

(4) **If** (Detect(T_4^{ON}) = "yes") **then**

    (4a) Append-head(T_{output}^S, z_n^1).

Else

    (4b) Append-head(T_{output}^S, z_n^0).

EndIf

(5) $T_0^S = \cup(T_3, T_4^{ON}, T_4^{OFF})$.

EndAlgorithm

Consider that a bit pattern, $10(y_2^1 y_1^0)$, is transmitted from one location to another location. Tubes T_0^S with the result shown in Table 3.6, $T_{Input}^S = \{y_2^1 y_1^0\}$ and $T_{Output}^S = \varnothing$, and they are regarded as input tubes of **Algorithm 3.6**. Because the value for n is two, Steps (1a) through (1f) will be run two times. After the first execution of Steps (1a) and (1b) is implemented, tube $T_0^S = \varnothing$, tube $T_1^{ON} = \{z_2^0 z_1^1 z_0^0 y_2^1 y_1^1, z_2^1 z_1^1 z_0^0 y_2^0 y_1^1\}$, tube $T_1^{OFF} = \{z_2^1 z_1^0 z_0^0 y_2^1 y_1^0, z_2^0 z_1^0 z_0^0 y_2^0 y_1^0\}$, tube $T_{Input}^S = \varnothing$, tube $T_2^{ON} = \varnothing$ and $T_2^{OFF} = \{y_2^1 \, y_1^0\}$. Next, because a "no" from the first execution of Step (1c) is returned, after the first execution of Step (1e) is performed, tube $T_0^S = \{z_2^1 \, z_1^0 \, z_0^0 \, y_2^0 \, y_1^0, \, z_2^0 \, z_1^0 \, z_0^0 \, y_2^0 \, y_1^0\}$, tube $T_1^{OFF} = \varnothing$, tube $T_3 = \{z_2^0 \, z_1^1 \, z_0^0 \, y_2^1 \, y_1^1, \, z_2^1 \, z_1^1 \, z_0^0 \, y_2^0 \, y_1^1\}$ and tube $T_1^{ON} = \varnothing$. After the rest of operations in Step (1) are run, tube $T_{Input}^S = \{y_2^1 \, y_1^0\}$, tube $T_0^S = \{z_2^1 \, z_1^0 \, z_0^0 \, y_2^1 \, y_1^0\}$, tube $T_3 = \{z_2^0 \, z_1^1 \, z_0^0 \, y_2^1 \, y_1^1, \, z_2^1 \, z_1^1 \, z_0^0 \, y_2^0 \, y_1^1, \, z_2^0 \, z_1^0 \, z_0^0 \, y_2^0 \, y_1^0\}$, tube $T_1^{ON} = \varnothing$, tube $T_1^{OFF} = \varnothing$, tube $T_2^{ON} = \varnothing$ and tube $T_2^{OFF} = \varnothing$.

Then, the first execution of Steps (2) and (3) is performed, tube $T_4^{ON} = \{z_2^1 \, z_1^0 \, z_0^0 \, y_2^1 \, y_1^0\}$, tube $T_4^{OFF} = \varnothing$, tube $T_0^S = \varnothing$, tube $T_{Output}^S = \{y_2^1 \, y_1^0\}$ and tube $T_{Input}^S = \varnothing$. Because a "yes" from the first execution of Step (4) is returned, after the first execution of Step (4a) is implemented, tube $T_{Output}^S = \{z_2^1 \, y_2^1 \, y_1^0\}$. Next, the first execution of Step (5) is performed, tube $T_0^S = \{z_2^0 \, z_1^1 \, z_0^0 \, y_2^1 \, y_1^1, \, z_2^1 \, z_1^1 \, z_0^0 \, y_2^0 \, y_1^1, \, z_2^1 \, z_1^1 \, z_0^0 \, y_2^1 \, y_1^0, \, z_2^0 \, z_1^0 \, z_0^0 \, y_2^0 \, y_1^0\}$, tube $T_3 = \varnothing$, tube $T_4^{ON} = \varnothing$ and tube $T_4^{OFF} = \varnothing$. Therefore, the result is shown in Table 3.7. Lemma 3-6 is applied to prove correction of **Algorithm 3.6**.

Table 3.7 **Algorithm 3.6** generates the result

Tube	**Algorithm 3.6** generates the result
T_0^S	$\{z_2^0 \, z_1^1 \, z_0^0 \, y_2^1 \, y_1^1, \, z_2^1 \, z_1^1 \, z_0^0 \, y_2^0 \, y_1^1, \, z_2^1 \, z_1^1 \, z_0^0 \, y_2^1 \, y_1^0, \, z_2^0 \, z_1^0 \, z_0^0 \, y_2^0 \, y_1^0\}$
T_{Output}^S	$\{z_2^1 \, y_2^1 \, y_1^0\}$
T_{Input}^S	$\varnothing$

Lemma 3-6: **Algorithm 3.6** *can be applied to finish the function of a parity generator of n bits.*

Proof Refer to **Lemma 3-5**.

3.9 Introduction for the Parity Checker of Error-Detection Codes on Digital Communication

Algorithm 3.7 can be applied to replace logic circuits of a "parity-checker" in receiving end. One one-bit binary number, c_1, is used to represent a parity-error bit. The value "0" for c_1 is applied to represent occurrence of no error during transmitted period to any received message. On the other hand, the value "1" of c_1 is used to represent occurrence of errors during transmitted period to any received message. Tube T_0^R is regarded as an input tube of **Algorithm 3.7**. The second parameter, n, in **Algorithm 3.7** is the number of bits for received messages. In **Algorithm 3.7**, the third parameter, tube T_{Input}^R, is applied to store any message received. Similarly, in **Algorithm 3.7**, the fourth parameter, tube T_{Output}^R, is employed to store those received messages, in which each message includes the corresponding parity-error bit that indicates occurrence of no error for it. The fifth parameter, tube T_{Bad}^R, is used to store those received messages with the corresponding parity-error bit that indicates occurrence of errors for them.

Algorithm 3.7: ParityChecker(T_0^R, n, T_{Input}^R, T_{Output}^R, T_{Bad}^R)

(1) **For** $g = 1$ **to** n

    (1a) $T_1^{ON} = +(T_0^R, y_g^1)$ and $T_1^{OFF} = -(T_0^R, y_g^1)$.

    (1b) $T_2^{ON} = +(T_{Input}^R, y_g^1)$ and $T_2^{OFF} = -(T_{Input}^R, y_g^1)$.

    (1c) **If** (Detect(T_2^{ON}) = "yes") **then**

        (1d) $T_0^R = \cup(T_0^R, T_1^{ON})$ and $T_3 = \cup(T_3, T_1^{OFF})$.

    Else

        (1e) $T_0^R = \cup(T_0^R, T_1^{OFF})$ and $T_3 = \cup(T_3, T_1^{ON})$.

    EndIf

    (1f) $T_{Input}^R = \cup(T_2^{ON}, T_2^{OFF})$.

EndFor

(2) $T_4^{ON} = +(T_0^R, z_n^1)$ and $T_4^{OFF} = -(T_0^R, z_n^1)$.

(3) $T_2^{ON} = +(T_{Input}^R, z_n^1)$ and $T_2^{OFF} = -(T_{Input}^R, z_n^1)$.

(4) **If** (Detect(T_2^{ON}) = "yes") **then**

    (4a) **If** (Detect(T_4^{ON}) = "yes") **then**

        (4b) $T_{Output}^R = (T_{Output}^R, T_2^{ON})$.

        (4c) Append-head(T_{Output}^R, c_1^0).

    Else

        (4d) $T_{Bad}^R = \cup(T_{Bad}^R, T_2^{ON})$.

        (4e) Append-head(T_{Bad}^R, c_1^1).

    EndIf

    Else

    (4f) **If** (Detect(T_4^{ON}) = "yes") **then**

        (4g) $T_{Bad}^R = \cup(T_{Bad}^R, T_2^{OFF})$.

        (4h) Append-head(T_{Bad}^R, c_1^1).

    Else

        (4i) $T_{Output}^R = \cup(T_{Output}^R, T_2^{OFF})$.

        (4j) Append-head(T_{Output}^R, c_1^0).

    EndIf

    EndIf

(5) **If** (Detect(T_{Bad}^R) = "yes") **then**

    (5a) Read(T_{Bad}^R).

EndIf

EndAlgorithm

Consider that in receiving end a bit pattern, $11(y_2^1\ y_1^1)$, and the corresponding parity bit, z_2^1, are received. Tubes T_0^R with the result shown in Table 3.6, $T_{Input}^R = \{z_2^1\ y_2^1\ y_1^1\}$, $T_{Output}^R = \varnothing$, and $T_{Bad}^R = \varnothing$, and they are regarded as input tubes of **Algorithm 3.7**. Step (1) is the only loop and is mainly applied to find the corresponding parity bit for the received message. At the end of Step (1), tube $T_{Input}^R = \{z_2^1\ y_2^1\ y_1^1\}$, tube $T_0^R = \{z_2^0\ z_1^1\ z_0^0\ y_2^1\ y_1^1\}$, tube $T_3 = \{z_2^1\ z_1^1\ z_0^0\ y_2^1\ y_1^1,\ z_2^1\ z_1^1\ z_0^0\ y_2^0\ y_1^0,\ z_2^0\ z_1^0\ z_0^0\ y_2^0\ y_1^0\}$, tube $T_1^{ON} = \varnothing$, tube $T_1^{OFF} = \varnothing$, tube $T_2^{ON} = \varnothing$ and tube $T_2^{OFF} = \varnothing$. Then, after each execution for Steps (2) through (4j) is implemented, the result is shown in Table 3.8. After the execution for Steps (5) and (5a) is performed, it is indicated that there is occurrence of errors for the received message during transmitted period. Lemma 3-7 is used to demonstrate correction of **Algorithm 3.7**.

Table 3.8 **Algorithm 3.7** generates the result

Tube	**Algorithm 3.7** generates the result
T_0^R	$\{z_2^0 z_1^1 z_0^0 y_2^1 y_1^1,\ z_2^1 z_1^1 z_0^0 y_2^0 y_1^1,\ z_2^1 z_1^0 z_0^0 y_2^1 y_1^0,\ z_2^0 z_1^0 z_0^0 y_2^0 y_1^0\}$
T_{Output}^R	$\varnothing$
T_{Bad}^R	$\{c_1^1 z_2^1 y_2^1 y_1^1\}$
T_{Input}^R	$\varnothing$

Lemma 3-7: **Algorithm 3.7** *can be applied to perform the function of a parity checker of n bits.*

Proof Refer to **Lemma 3-5.**∎

3.10 Summary

In this chapter an introduction to how eight bio-molecular operations were applied to complete representation of bit patterns for data stored in tubes and to deal with various problems was provided. We described a set P with $\{x_n \ldots x_1|$ each x_k was a binary value for $1 \leq k \leq n$, where $n \geq 0\}$ and a tube that was a storage device in which data represented by bit patterns were stored. We then introduced that a set P could be regarded as a tube T and an element in the set P could also be regarded as one data stored in the tube T. We used **Definition 3-1** to show the function of the *append* operation, and we also applied **Algorithm 3.1** and its proof to explain how the *append* operation constructed one bit pattern, $x_n \ldots x_1$, that was stored in a tube T. We then applied **Definition 3-2** to reveal the function of the *amplify* operation, and we also used **Algorithm 3.2** and its proof to demonstrate how the *amplify* operation generated the same two bit patterns, $x_n \ldots x_1$.

The function of the *merge* operation was introduced from **Definition 3-3**, and **Algorithm 3.3** and its proof were applied to show how the *merge* operation and other two operations constructed 2^n combinational states of n bits. Next, the function of the *extract* operation, the *detect* operation, the *discard* operation, the *append-head* operation and the *read* operation was described from **Definitions 3-4** through **3-8**. A one-bit parity counter is to count whether the number of 1's for two input bits is odd or even or not. **Algorithm 3.4** and its proof were used to explain how the function of a one-bit parity counter was implemented by means of eight bio-molecular operations above.

A parity counter of n bits is to count whether the number of 1's for 2^n combinational states is odd or even by means of n times of this one-bit parity counter.

Algorithm 3.5 and its proof were applied to reveal how the function of a parity counter of n bits was implemented by means of eight bio-molecular operations. For digital computer systems, during transfer of information from one location to another location, in sending end a "parity-generation" is used to generate the corresponding parity bit for it and in receiving end a "parity-checker" is applied to check the proper parity adopted. **Algorithm 3.6** and its proof were applied to show how logic circuits of a "parity-generator" in sending end were implemented by means of eight bio-molecular operations, and **Algorithm 3.7** and its proof were employed to demonstrate how logic circuits of a "parity-checker" in receiving end were implemented by means of eight bio-molecular operations.

3.11 Bibliographical Notes

In this chapter, for a more detailed introduction to three constructs and flowcharts of sequence, decision and repetition, the recommended book are Bjorner (2006); Forouzan and Mosharraf (2008); James and Witold (2000); Reed (2008). The first chapter and the second chapter of the book by Paun et al. (Paun et al. 1998) is a concise introduction to physical implementation of bio-molecular operations. The book by Amos (Amos 2005) is a more detailed description to physical implementation of bio-molecular operations. The book by Drlica (Drlica 1992) is a good introduction for molecular biology and genetic engineering for a general reader without biochemistry and biology. The book by Mano (Mano 1979) is a beautiful introduction to digital logic circuits. The two articles in Nakano et al. (2012); Nakano et al. (2013) are very good introduction to engineered biological nanomachines to communicate with biological systems at the molecular level. The article in Felicetti et al. (2014) is a good illustration for a communication protocol between biological nanomachines built upon molecular communications. The article in Nakano et al. (2014) is a good introduction for applying the layered architectural approach, traditionally used in computer networks, to the design and development of molecular communication systems of biological nanomachines.

3.12 Exercises

3.1 The truth table of a logical operation **NOT** with one input and one output is shown in Table 3.9. Based on Table 3.9, write a bio-molecular program to implement the function of the logical operation **NOT**.

Table 3.9 The truth table of a logical operation **NOT** is shown

The first input	The first output
0	1
1	0

3.2 The truth table of a logical operation **AND** with two inputs and one output is shown in Table 3.10. Based on Table 3.10, write a bio-molecular program to implement the function of the logical operation **AND**.

Table 3.10 The truth table of a logical operation **AND** is shown

The first input	The second input	The first output
0	0	0
0	1	0
1	0	0
1	1	1

3.3 The truth table of a logical operation **OR** with two inputs and one output is shown in Table 3.11. Based on Table 3.11, write a bio-molecular program to implement the function of the logical operation **OR**.

Table 3.11 The truth table of a logical operation **OR** is shown

The first input	The second input	The first output
0	0	0
0	1	1
1	0	1
1	1	1

3.4 The truth table of a logical operation **BUFFER** with one input and one output is shown in Table 3.12. Based on Table 3.12, write a bio-molecular program to implement the function of the logical operation **BUFFER**.

Table 3.12 The truth table of a logical operation **BUFFER** is shown

The first input	The first output
0	0
1	1

3.5 The truth table of a logical operation **NAND** with two inputs and one output is shown in Table 3.13. Based on Table 3.13, write a bio-molecular program to implement the function of the logical operation **NAND**.

Table 3.13 The truth table of a logical operation **NAND** is shown

The first input	The second input	The first output
0	0	1
0	1	1
1	0	1
1	1	0

3.6 The truth table of a logical operation **NOR** with two inputs and one output is shown in Table 3.14. Based on Table 3.14, write a bio-molecular program to implement the function of the logical operation **NOR**.

Table 3.14 The truth table of a logical operation **NOR** is shown	The first input	The second input	The first output
	0	0	1
	0	1	0
	1	0	0
	1	1	0

3.7 The truth table of a logical operation **Exclusive-OR** with two inputs and one output is shown in Table 3.15. Based on Table 3.15, write a bio-molecular program to implement the function of the logical operation **Exclusive-OR**.

Table 3.15 The truth table of a logical operation **Exclusive-OR** is shown	The first input	The second input	The first output
	0	0	0
	0	1	1
	1	0	1
	1	1	0

3.8 The truth table of a logical operation **Exclusive-NOR** with two inputs and one output is shown in Table 3.16. Based on Table 3.16, write a bio-molecular program to implement the function of the logical operation **Exclusive-NOR**.

Table 3.16 The truth table of a logical operation **Exclusive-NOR** is shown	The first input	The second input	The first output
	0	0	1
	0	1	0
	1	0	0
	1	1	1

3.9 The truth table of a logical operation **NULL** with two inputs and one output is shown in 3.17. Based on Table 3.17, write a bio-molecular program to implement the function of the logical operation **NULL**.

Table 3.17 The truth table of a logical operation **NULL** is shown

The first input	The second input	The first output
0	0	0
0	1	0
1	0	0
1	1	0

3.10 The truth table of a logical operation **IDENTITY** with two inputs and one output is shown in Table 3.18. Based on Table 3.18, write a bio-molecular program to implement the function of the logical operation **IDENTITY**.

Table 3.18 The truth table of a logical operation **IDENTITY** is shown

The first input	The second input	The first output
0	0	1
0	1	1
1	0	1
1	1	1

References

D. Bjorner, *Software engineering 3*: *Domains, Requirements, and Software Design*. (Springer, Heidelberg, 2006). ISBN: 9783540211518

B. Forouzan, F. Mosharraf, *Foundations of Computer Science*, 2nd edn. (Thomson, London, 2008). ISBN: 978-1-84480-700-0

P.F. James, P. Witold, *Software Engineering*: *An Engineering Approach*. (John Wiley, Boca Raton, 2000). ISBN: 0471189642

D. Reed, *A Balanced Introduction to Computer Science*. (Pearson Prentice Hall, New Jerst, 2008). ISBN: 9780136017226

G. Paun, G. Rozenberg, A. Salomaa, *DNA Computing: New Computing Paradigms*. (Springer, Hidelberg, 1998). ISBN: 3540641963

M. Amos, *Theoretical and Experimental DNA Computation*. (Springer, Hidelberg, 2005). ISBN: 9783540657736

K. Drlica, *Understanding DNA and Gene Clonig*: *A Guide for the CURIOUS*. (John Wiley and Sons, Boca Raton, 1992). ISBN: 9780471434160

M. M. Mano, *Digital Logic and Computer Design*. (Prentice-Hall, New Jersy, 1979). ISBN: 0-13-214510-3

T. Nakano, M.J. Moore, F. Wei, A.V. Vasilakos, J. Shuai, Molecular communication and networking: opportunities and challenges. IEEE Trans. Nanobiosci. **11**(2), 135–148 (2012)

T. Nakano, Y. Okaie, A.V. Vasilakos, Transmission rate control for molecular communication among biological nanomachines. IEEE J. Sel. Areas Commun. **31**(10), 1–12 (2013)

T. Nakano, T. Suda, Y. Okaie, M. J. Moore, A. V. Vasilakos, A layered architecture approach for molecular communication among biological nanomachines. IEEE Trans. Nanobiosci. (2014)

L. Felicetti, M. Femminella, G. Reali, T. Nakano, A.V. Vasilakos, TCP-like molecular communications. IEEE J. Sel. Areas Comm. (2014)

Chapter 4
Introduction for Number Representation on Bio-molecular Computer

In this chapter we first introduce two numbering systems: the decimal system the binary system. Next we describe how to convert a number from the decimal system to the binary system vice versa. Finally we introduce how numbers in the form of bit patterns are stored inside a tube in bio-molecular computer.

4.1 Introduction to Decimal and Binary

Today, the decimal system and the binary system are the most popular two numbering systems. The world currently uses the decimal system to numbers developed by Arabian mathematicians in the eighth century. The first people for using a decimal numbering system were the ancient Egyptians. The Babylonians enhanced on the Egyptian system by making the positions in the decimal numbering system meaningful.

The base of the decimal system is 10. For the decimal system, the first position is 10 raised to the power 0, the second position is 10 raised to the power 1 and the nth position is 10 raised to the power n. The relationship between the powers and the number 128 is shown in Fig. 4.1.

Fig. 4.1 Decimal system

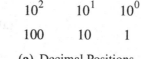

$$10^2 \quad 10^1 \quad 10^0$$
$$100 \quad 10 \quad 1$$

(a) Decimal Positions

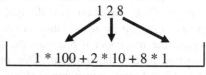

$$1 * 100 + 2 * 10 + 8 * 1$$

(b) Decimal Representation

W.-L. Chang and A. V. Vasilakos, *Molecular Computing*, Studies in Big Data 4,
DOI: 10.1007/978-3-319-05122-2_4, © Springer International Publishing Switzerland 2014

Whereas the decimal system is based on 10, the binary system is based on 2. There are only two digits in the binary system, 0 and 1. The positional weights for a binary system and the value 128 in binary are shown in Fig. 4.2. In the position table, each position is double the previous position. Again, this is because the base of the system is 2.

Fig. 4.2 Binary system

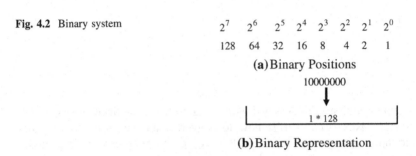

2^7 2^6 2^5 2^4 2^3 2^2 2^1 2^0

128 64 32 16 8 4 2 1

(a) Binary Positions

10000000

1 * 128

(b) Binary Representation

4.2 Conversion for Between Decimal and Binary

We begin by converting a number from the binary system to the decimal system. Start with the binary number and multiply each binary digit by its weight. Because each binary bit can be only 0 or 1, the result will be either 0 or the value of the weight. After multiplying all the digits, add the results. Binary to decimal conversion is shown in Fig. 4.3.

Fig. 4.3 Binary to decimal conversion

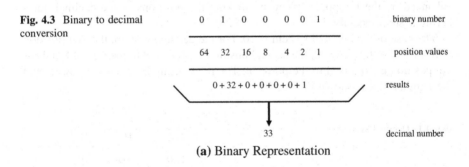

| 0 | 1 | 0 | 0 | 0 | 0 | 1 | binary number |

| 64 | 32 | 16 | 8 | 4 | 2 | 1 | position values |

0 + 32 + 0 + 0 + 0 + 0 + 1   results

33   decimal number

(a) Binary Representation

We use repetitive division for converting from decimal to binary. The original number, 33, in the example is divided by 2. The remainder, 1, becomes the first binary digit, and the second digit is determined by dividing the quotient, 16, by 2. Again, the remainder 0 becomes the binary digit, and 2 to determine the next position divides the quotient. This process continues until the quotient is 0. Decimal to binary conversion is shown in Fig. 4.4.

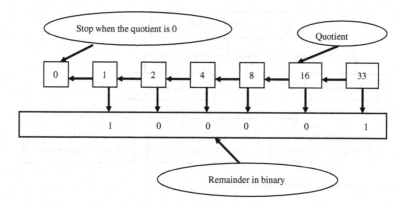

Fig. 4.4 Decimal to binary conversion

4.3 Integer Representation on Bio-molecular Computer

Integers are whole numbers (i.e., numbers without a fraction). For example, 33 is an integer, but 33.33 is not. As another example, −33 is an integer, but −33.33 is not. An integer can be positive or negative. A negative integer ranges from negative infinity to 0; a positive integer ranges from 0 to positive infinity (Fig. 4.5). All the integers in this range are represented with an infinity number of bits. This implies no bio-molecular computer with infinite storage capability.

Fig. 4.5 Range of integers

On a traditional computer, in order to use memory more efficiently, two broad categories of integer representation have been developed: unsigned integers and signed integers. Signed integers may also be represented in three different ways (Fig. 4.6). Because those ways for representing integers are broadly used, on bio-molecular computer the same ways are applied to represent integers. Note that today two's complement is the most commonly used representation of integers. However, the other representations are simpler and serve as a good foundation for two's complement, so they are first discussed in next subsections.

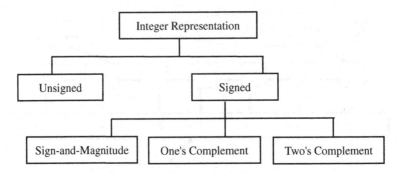

Fig. 4.6 Representation formats of integers

4.3.1 Introduction for Unsigned Integer Format

An unsigned integer is an integer without a sign. Its range is between 0 and positive infinity. However, on a traditional computer, no memory enough can be applied to represent all the integers in this range. Therefore, a constant defined on a traditional computer is called the maximum unsigned integer. An unsigned integer ranges between 0 and this constant. The maximum unsigned integer depends on the number of bits the traditional computer allocates to store an unsigned integer. From the statements above, in bio-molecular computer, the following definition defines the range of an unsigned integer.

Definition 4-1: Suppose that an n-bit binary number, $x_n \ldots x_1$ is used to represent an unsigned integer of n bits, where the value of each bit x_k is either 1 or 0 for $1 \leq k \leq n$. The bits x_n and x_1 is applied to represent, respectively, the most significant bit and the least significant bit for an unsigned integer of n bits. An unsigned integer of n bits ranges between 0 and $2^n - 1$.

Algorithm 4.1 is employed to construct the range of the value for an unsigned integer of n bits. Tube T_0 is an empty tube, and it is regarded as one input tube of **Algorithm 4.1**. The second parameter in **Algorithm 4.1**, n, is applied to represent the number of bits for an unsigned integer.

> **Algorithm 4.1: Generate-Unsigned-Integers(T_0, n)**
> (0a) Append-head(T_1, x_1^1).
> (0b) Append-head(T_2, x_1^0).
> (0c) $T_0 = \cup(T_1, T_2)$.
> (1) **For** $k = 2$ **to** n
>   (1a) Amplify(T_0, T_1, T_2).
>   (1b) Append-head(T_1, x_k^1).
>   (1c) Append-head(T_2, x_k^0).
>   (1d) $T_0 = \cup(T_1, T_2)$.
> **EndFor**
> **EndAlgorithm**

Consider that eight values for an unsigned integer of three bits are, respectively, $000(0_{10})$, $001(1_{10})$, $010(2_{10})$, $011(3_{10})$, $100(4_{10})$, $101(5_{10})$, $110(6_{10})$ and $111(7_{10})$. Tube T_0 is an empty tube and is regarded as an input tube of **Algorithm 4.1**. The second parameter, n, in **Algorithm 4.1** is the number of bits for representing an unsigned integer and its value is three. After the first execution for Step (0a) and the first execution for Step (0b) are performed, tube $T_1 = \{x_1^1\}$ and tube $T_2 = \{x_1^0\}$. Then, after the first execution of Step (0c) is implemented, tube $T_0 = \{x_1^1, x_1^0\}$, tube $T_1 = \varnothing$ and tube $T_2 = \varnothing$.

Step (1) is the main loop and the lower bound and the upper bound are, respectively, two and three, so Steps (1a) through (1d) will be run two times. After the first execution of Step (1a) is finished, tube $T_0 = \varnothing$, tube $T_1 = \{x_1^1, x_1^0\}$ and tube $T_2 = \{x_1^1, x_1^0\}$. Next, after the first execution for Step (1b) and Step (1c) is performed, tube $T_1 = \{x_2^1 \, x_1^1, x_2^1 \, x_1^0\}$ and tube $T_2 = \{x_2^0 \, x_1^1, x_2^0 \, x_1^0\}$. After the first execution of Step (1d) is implemented, tube $T_0 = \{x_2^1 \, x_1^1, x_2^1 \, x_1^0, x_2^0 \, x_1^1, x_2^0 \, x_1^0\}$, tube $T_1 = \varnothing$ and tube $T_2 = \varnothing$.

Then, after the second execution of Step (1a) is finished, tube $T_0 = \varnothing$, tube $T_1 = \{x_2^1 \, x_1^1, x_2^1 \, x_1^0, x_2^0 \, x_1^1, x_2^0 \, x_1^0\}$ and tube $T_2 = \{x_2^1 \, x_1^1, x_2^1 \, x_1^0, x_2^0 \, x_1^1, x_2^0 \, x_1^0\}$. After the rest of operations are performed, tube $T_1 = \varnothing$, tube $T_2 = \varnothing$ and the result for tube T_0 is shown in Table 4.1. **Lemma 4-1** is applied to demonstrate correction of **Algorithm 4.1**.

	Tube	The result is generated by **Algorithm 4.1**
Table 4.1 The result for tube T_0 is generated by **Algorithm 4.1**	T_0	$\{x_3^1 \, x_2^1 \, x_1^1, x_3^1 \, x_2^1 \, x_1^0, x_3^1 \, x_2^0 \, x_1^1, x_3^1 \, x_2^0 \, x_1^0,$ $x_3^0 \, x_2^1 \, x_1^1, x_3^0 \, x_2^1 \, x_1^0, x_3^0 \, x_2^0 \, x_1^1, x_3^0 \, x_2^0 \, x_1^0\}$

Lemma 4-1: **Algorithm 4.1** *can be used to construct the range of the value for an unsigned integer of n bits.*

Proof An unsigned integer of n bits, $x_n \dots x_1$, ranges between 0 and $2^n - 1$. This implies that the domain of the value is a combination of 2^n binary states and is actually equal to the Cartesian production of each bit, $\{x_n \dots x_1 | x_k \in \{0, 1\}$ for $1 \leq k \leq n\}$. **Algorithm 4.1** is implemented by means of the *extract, append-head, amplify,* and *merge* operations. Each execution of Step (0a) and each execution of Step (0b), respectively, append the value "1" for x_1 as the first bit of every data stored in tube T_1 and the value "0" for x_1 as the first bit of every data stored in tube T_2. Next, each execution of Step (0c) is to pour tubes T_1 and T_2 into tube T_0. This implies that tube T_0 contains all of the data that have $x_1 = 1$ and $x_1 = 0$ and tubes T_1 and T_2 become empty tubes.

Step (1) is the only loop and is mainly used to perform the Cartesian production of each bit. On the first execution of Step (1a), it is used to amplify tube T_0 and to generate two new tubes, T_1 and T_2, which are copies of T_0. Tube T_0 then becomes empty. Then, the first execution of Step (1b) is applied to append the value "1" for x_2 onto tube T_1. This is to say that those unsigned integers containing the value "1" to the *second* bit appear in tube T_1. On the first execution of Step (1c), it is also employed to append the value "0" for x_2 onto tube T_2. That implies that these unsigned integers containing the value "0" to the *second* bit appear in tube T_2. Next, the first execution of Step (1d) is used to pour tubes T_1 and T_2 into tube T_0. This implies that the Cartesian production to the second bit is generated and stored in tube T_0. Repeat to execute Steps (1a) through (1d) until the nth bit is processed. The Cartesian production of n bits is generated and stored in tube T_0. Therefore, it is inferred that **Algorithm 4.1** can be used to construct the range of the value for an unsigned integer of n bits. ■

4.3.2 Introduction for Sign-and-Magnitude Integer Format

In the sign-and magnitude format, storing an integer requires a bit to represent the sign. Generally speaking, 0 is applied to represent the positive and 1 is employed to represent the negative. This implies that in n-bit allocation, only $(n-1)$ bits can be applied to represent the absolute value of the number (number without the sign). Hence, the maximum positive value is one half the unsigned value. From the statements above, in bio-molecular computer, the following definition defines the range of a sign-and-magnitude integer.

Definition 4-2: Suppose that an n-bit binary number, $x_n \ldots x_1$ is used to represent a sign-and-magnitude integer of n bits, where the value of each bit x_k is either 1 or 0 for $1 \leq k \leq n$. The bit x_n is used to represent the sign, and the bits x_{n-1} and x_1 is applied to represent, respectively, the most significant bit and the least significant bit for a sign-and-magnitude integer of n bits. A sign-and-magnitude integer of n bits ranges between $-(2^{n-1} - 1)$ and $+(2^{n-1} - 1)$.

From **Definition 4-2**, it is very clear that there are two 0s in sign-and-magnitude representation: positive and negative. For example, in an 8-bit allocation: "00000000" is applied to represent "+0" and "10000000" is used to represent "−0". **Algorithm 4.2** is applied to construct the range of the value for a sign-and-magnitude integer of n bits. Tube T_0 is an empty tube, and it is regarded as one input tube of **Algorithm 4.2**. The second parameter in **Algorithm 4.2**, n, is applied to represent the number of bits for a sign-and-magnitude integer.

Algorithm 4.2: Generate-Sign-and-Magtitude-Integers(T_0, n)
(0a) Append-head(T_1, x_1^1).
(0b) Append-head(T_2, x_1^0).
(0c) $T_0 = \cup(T_1, T_2)$.
(1) **For** $k = 2$ **to** n
    (1a) Amplify(T_0, T_1, T_2).
    (1b) Append-head(T_1, x_k^1).
    (1c) Append-head(T_2, x_k^0).
    (1d) $T_0 = \cup(T_1, T_2)$.
EndFor
EndAlgorithm

Consider that eight values for a sign-and-magnitude integer of three bits are, respectively, $000(+0_{10})$, $001(+1_{10})$, $010(+2_{10})$, $011(+3_{10})$, $100(-0_{10})$, $101(-1_{10})$, $110(-2_{10})$ and $111(-3_{10})$. Tube T_0 is an empty tube and is regarded as an input tube of **Algorithm 4.2**. The second parameter, n, in **Algorithm 4.2** is the number of bits for representing a sign-and-magnitude integer and its value is three. After the first execution of Step (0a) and the first execution of Step (0b) are performed, tube $T_1 = \{x_1^1\}$ and tube $T_2 = \{x_1^0\}$. Then, after the first execution of Step (0c) is implemented, tube $T_0 = \{x_1^1, x_1^0\}$, tube $T_1 = \varnothing$ and tube $T_2 = \varnothing$.

Step (1) is the main loop and the lower bound and the upper bound are, subsequently, are two and three, so Steps (1a) through (1d) will be run two times. After the first execution of Step (1a) is finished, tube $T_0 = \varnothing$, tube $T_1 = \{x_1^1, x_1^0\}$ and tube $T_2 = \{x_1^1, x_1^0\}$. Next, after the first execution for Step (1b) and Step (1c) is performed, tube $T_1 = \{x_2^1 x_1^1, x_2^1 x_1^0\}$ and tube $T_2 = \{x_2^0 x_1^1, x_2^0 x_1^0\}$. After the first execution of Step (1d) is implemented, tube $T_0 = \{x_2^1 x_1^1, x_2^1 x_1^0, x_2^0 x_1^1, x_2^0 x_1^0\}$, tube $T_1 = \varnothing$ and tube $T_2 = \varnothing$.

Then, after the second execution of Step (1a) is finished, tube $T_0 = \varnothing$, tube $T_1 = \{x_2^1 x_1^1, x_2^1 x_1^0, x_2^0 x_1^1, x_2^0 x_1^0\}$ and tube $T_2 = \{x_2^1 x_1^1, x_2^1 x_1^0, x_2^0 x_1^1, x_2^0 x_1^0\}$. After the rest of operations are performed, tube $T_1 = \varnothing$, tube $T_2 = \varnothing$ and the result for tube T_0 is shown in Table 4.2. **Lemma 4-2** is applied to prove correction of **Algorithm 4.2**.

Table 4.2 The result for tube T_0 is generated by **Algorithm 4.2**

Tube	The result is generated by **Algorithm 4.2**
T_0	$\{x_3^1 x_2^1 x_1^1, x_3^1 x_2^1 x_1^0, x_3^1 x_2^0 x_1^1, x_3^1 x_2^0 x_1^0,$ $x_3^0 x_2^1 x_1^1, x_3^0 x_2^1 x_1^0, x_3^0 x_2^0 x_1^1, x_3^0 x_2^0 x_1^0\}$

Lemma 4-2: **Algorithm 4.2** *can be used to construct the range of the value for a sign-and-magnitude integer of n bits.*

Proof Refer to **Lemma 4-1**.

4.3.3 Introduction for One's Complement Integer Format

In the one's complement format, a different convention is adopted. Representing a positive number uses the convention adopted for an unsigned integer. To represent a negative number complements the positive number. In other words, $+15$ is represented just like an unsigned integer, and -15 is represented as the complement of $+15$. In one's complement, the complement of a number is obtained by means of changing all 1s to 0s and all 0s to 1s. From the statements above, in bio-molecular computer, the following definition defines the range of a one's complement integer.

Definition 4-3: Suppose that an n-bit binary number, $x_n \ldots x_1$ is used to represent a one's complement integer of n bits, where the value of each bit x_k is either 1 or 0 for $1 \leq k \leq n$. A one's complement integer of n bits ranges between $-(2^{n-1}-1)$ and $+(2^{n-1}-1)$.

From **Definition 4-3**, it is indicated that there are two 0s in one's complement representation: positive and negative. For example, in an 8-bit allocation: "00000000" is applied to represent "+0" and "11111111" is used to represent "–0". **Algorithm 4.3** is applied to construct the range of the value for a one's complement integer of n bits. Tube T_0 is an empty tube, and it is regarded as one input tube of **Algorithm 4.3**. The second parameter in **Algorithm 4.3**, n, is applied to represent the number of bits for a one's complement integer.

Algorithm 4.3: Generate-One's-Complement-Integers(T_0, n)

(0a) Append-head($T_1, x_1{}^1$).

(0b) Append-head($T_2, x_1{}^0$).

(0c) $T_0 = \cup(T_1, T_2)$.

(1) **For** $k = 2$ **to** n

    (1a) Amplify(T_0, T_1, T_2).

    (1b) Append-head($T_1, x_k{}^1$).

    (1c) Append-head($T_2, x_k{}^0$).

    (1d) $T_0 = \cup(T_1, T_2)$.

EndFor

EndAlgorithm

Consider that eight values for a one's complement integer of three bits are, respectively, $000(+0_{10})$, $001(+1_{10})$, $010(+2_{10})$, $011(+3_{10})$, $100(-3_{10})$, $101(-2_{10})$, $110(-1_{10})$ and $111(-0_{10})$. Tube T_0 is an empty tube and is regarded as an input tube of **Algorithm 4.3**. The second parameter, n, in **Algorithm 4.3** is the number of bits for representing a one's complement integer and its value is three. After the first execution of Step (0a) and the first execution of Step (0b) are performed, tube $T_1 = \{x_1^1\}$ and tube $T_2 = \{x_1^0\}$. Then, after the first execution of Step (0c) is implemented, tube $T_0 = \{x_1^1, x_1^0\}$, tube $T_1 = \varnothing$ and tube $T_2 = \varnothing$.

Step (1) is the main loop and the lower bound and the upper bound are, respectively, two and three, so Steps (1a) through (1d) will be run two times. After the first execution of Step (1a) is finished, tube $T_0 = \varnothing$, tube $T_1 = \{x_1^1, x_1^0\}$ and tube $T_2 = \{x_1^1, x_1{}^0\}$. Next, after the first execution for Step (1b) and Step (1c) is performed, tube $T_1 = \{x_2^1 x_1^1, x_2{}^1 x_1^0\}$ and tube $T_2 = \{x_2^0 x_1^1, x_2{}^0 x_1^0\}$. After the first execution of Step (1d) is implemented, tube $T_0 = \{x_2^1 x_1^1, x_2{}^1 x_1^0, x_2^0 x_1^1, x_2{}^0 x_1^0\}$, tube $T_1 = \varnothing$ and tube $T_2 = \varnothing$.

Then, after the second execution of Step (1a) is finished, tube $T_0 = \varnothing$, tube $T_1 = \{x_2^1 x_1^1, x_2{}^1 x_1^0, x_2^0 x_1^1, x_2{}^0 x_1^0\}$ and tube $T_2 = \{x_2^1 x_1^1, x_2{}^1 x_1^0, x_2^0 x_1^1, x_2{}^0 x_1^0\}$. After the rest of operations are performed, tube $T_1 = \varnothing$, tube $T_2 = \varnothing$ and the result for tube T_0 is shown in Table 4.3. **Lemma 4-3** is applied to prove correction of **Algorithm 4.3**.

Table 4.3 The result for tube T_0 is generated by Algorithm 4.3	Tube	The result is generated by Algorithm 4.3
	T_0	$\{x_3^1 \, x_2^1 \, x_1^1, \, x_3^1 \, x_2^1 \, x_1^0, \, x_3^1 \, x_2^0 \, x_1^1, \, x_3^1 \, x_2^0 \, x_1^0,$
		$x_3^0 \, x_2^1 \, x_1^1, \, x_3^0 \, x_2^1 \, x_1^0, \, x_3^0 \, x_2^0 \, x_1^1, \, x_3^0 \, x_2^0 \, x_1^0\}$

Lemma 4-3: **Algorithm 4.3** *can be used to construct the range of the value for a one's complement integer of n bits.*

Proof Refer to **Lemma 4-1**.

4.3.4 Introduction for Two's Complement Integer Format

Because previously mentioned in Sect. 4.3.3, two 0s (+0 and −0) are represented in one's complement. This can yield some confusion in computations. For example, if you add a number and its complement (+3 and −3) in one's complement, you obtain negative −0 instead of +0. Two's complement representation can be applied to solve all those problems.

In two's complement representation, representing a positive number uses the convention adopted for a sign-and-magnitude integer. To represent a negative number is to take the two's complement of the positive number, including its sign bit. This is to say that the first step is to obtain the one's complement of the positive number by means of changing all 1s to 0s and all 0s to 1s and then the second step is to add one to the one's complement of the positive number. For example, +3 is represented as 011 in three bits, and −3 is represented as 101 in three bits in two's complement. From the statements above, in bio-molecular computer, the following definition is applied to define the range of a two's complement integer.

Definition 4-4: Suppose that an n-bit binary number, $x_n \ldots x_1$ is used to represent a two's complement integer of n bits, where the value of each bit x_k is either 1 or 0 for $1 \le k \le n$. A two's complement integer of n bits ranges between $-(2^{n-1})$ and $+(2^{n-1}-1)$.

From **Definition 4-4**, it is pointed out that there is only one 0 in two's complement representation. For example, in an 8-bit allocation: "00000000" is used to represent "0". **Algorithm 4.4** is employed to construct the range of the value for a two's complement integer of n bits. Tube T_0 is an empty tube, and it is regarded as one input tube of **Algorithm 4.4**. The second parameter in **Algorithm 4.4**, n, is employed to represent the number of bits for a two's complement integer.

Algorithm 4.4: Generate-Two's-Complement-Integers(T_0, n)

(0a) Append-head(T_1, x_1^1).

(0b) Append-head(T_2, x_1^0).

(0c) $T_0 = \cup(T_1, T_2)$.

(1) **For** $k = 2$ **to** n

    (1a) Amplify(T_0, T_1, T_2).

    (1b) Append-head(T_1, x_k^1).

    (1c) Append-head(T_2, x_k^0).

    (1d) $T_0 = \cup(T_1, T_2)$.

EndFor

EndAlgorithm

Consider that eight values for a two's complement integer of three bits are, respectively, $000(+0_{10})$, $001(+1_{10})$, $010(+2_{10})$, $011(+3_{10})$, $100(-4_{10})$, $101(-3_{10})$, $110(-2_{10})$ and $111(-1_{10})$. Tube T_0 is an empty tube and is regarded as an input tube of **Algorithm 4.4**. The second parameter in **Algorithm 4.4**, n, is employed to represent the number of bits for a two's complement integer and its value is three. After the first execution for Step (0a) and Step (0b) is implemented, tube $T_1 = \{x_1^1\}$ and tube $T_2 = \{x_1^0\}$. Next, after the first execution of Step (0c) is implemented, tube $T_0 = \{x_1^1, x_1^0\}$, tube $T_1 = \varnothing$ and tube $T_2 = \varnothing$.

Step (1) is the main loop and its lower and upper bounds are, respectively, two and three, so Steps (1a) through (1d) will be run two times. After the first execution of Step (1a) is performed, tube $T_0 = \varnothing$, tube $T_1 = \{x_1^1, x_1^0\}$ and tube $T_2 = \{x_1^1, x_1^0\}$. Next, after the first execution for Step (1b) and Step (1c) is implemented, tube $T_1 = \{x_2^1 x_1^1, x_2^1 x_1^0\}$ and tube $T_2 = \{x_2^0 x_1^1, x_2^0 x_1^0\}$. After the first execution of Step (1d) is implemented, tube $T_0 = \{x_2^1 x_1^1, x_2^1 x_1^0, x_2^0 x_1^1, x_2^0 x_1^0\}$, tube $T_1 = \varnothing$ and tube $T_2 = \varnothing$. Then, after the second execution of Step (1a) is performed, tube $T_0 = \varnothing$, tube $T_1 = \{x_2^1 x_1^1, x_2^1 x_1^0, x_2^0 x_1^1, x_2^0 x_1^0\}$ and tube $T_2 = \{x_2^1 x_1^1, x_2^1 x_1^0, x_2^0 x_1^1, x_2^0 x_1^0\}$. After the rest of operations are performed, tube $T_1 = \varnothing$, tube $T_2 = \varnothing$ and the result for tube T_0 is shown in Table 4.4. **Lemma 4-4** is used to show correction of **Algorithm 4.4**.

Table 4.4 The result for tube T_0 is generated by **Algorithm 4.4**	Tube	The result is generated by **Algorithm 4.4**
	T_0	$\{x_3^1 x_2^1 x_1^1, x_3^1 x_2^1 x_1^0, x_3^1 x_2^0 x_1^1, x_3^1 x_2^0 x_1^0,$ $x_3^0 x_2^1 x_1^1, x_3^0 x_2^1 x_1^0, x_3^0 x_2^0 x_1^1, x_3^0 x_2^0 x_1^0\}$

Lemma 4-4: **Algorithm 4.4** *can be applied to construct the range of the value for a two's complement integer of n bits.*

Proof Refer to **Lemma 4-1**.

4.4   Introduction for Floating-Point Representation

To represent a *floating-point* number (a number including an integer and a *fraction*), the number is divided into two parts. The first part is the integer and the second part is the fraction. For example, for a floating-point number (1.1_{10}), its integer part and fraction part are, respectively, 1_{10} and 0.1_{10}. To convert a floating-point number to a binary number contains three steps. The first step is to convert the integer part to binary. Next, the second step is to convert the fraction part to binary. The third step is to put a decimal point between the two parts. The procedure for finishing the first step is the same as that proposed in Sect. 4.2. For the second step, repetitive multiplication is applied to perform the task. For example, to convert 1.5 to binary, 2 multiply the fraction (0.5), and the result is 1.0. The integer part of the result (1) is extracted and becomes the leftmost binary digit. Because the fraction part of the result becomes "0", the converting task is performed. Figure 4.7 is employed to explain the process.

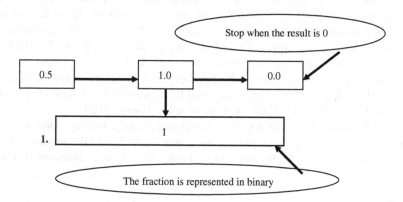

Fig. 4.7   Fraction to binary conversion

Table 4.5   Examples of normalization

An original floating-point number	Move	Normalized
$+10.1_2$	1 (the left moving)	$+2^1 \times 1.01$
-0.101_2	1 (the right moving)	$-2^{-1} \times 1.01$

After a floating-point number is normalized, three pieces of information about the floating-point number are stored. They are, respectively, sign, exponent, and mantissa (the bits of the right of the decimal point). For example, $+10.1_2$ in Table 4.5 becomes $+2^1 \times 1.01$ after it is normalized. For $+2^1 \times 1.01$, the sign is "+", the exponent is "1", and the mantissa is "01". Note that the 1 for the left of the decimal point is not stored and it is easily understood. One bit (0 or 1) can be applied to denote the sign, where 0 is used to represent the positive number and 1 is employed to represent the negative number. The exponent (power of 2) is applied to denote the movement of the decimal point. The mantissa is the binary number to the right of the decimal point. It is used to define the precision of a floating-point number. The mantissa is stored as an unsigned integer. The Institute of Electrical and Electronics Engineers (IEEE) had defined single precision and double precision for storing floating-point numbers. These formats will be described in the following subsections.

4.4.1 Introduction for Single Precision of Floating-Point Numbers

In Turing's machine, the system for storing the exponential value of a floating-point number is called the Excess system. In this system, to transform a floating-point number from decimal to binary or from binary to decimal is easy. In an Excess system, a positive number, called the *magic number*, is applied in the conversion process. The magic number is normally $(2^{n-1}-1)$ or (2^{n-1}), where n is the bit allocation. For example, if n is 8, the magic number is 127 or 128. The first case, 127, is called the representation Excess_127, and the second case, 128, is called the representation Excess_128. Similarly, if n is 11, the magic number is 1023 or 1024. The first case, 1023, is called the representation Excess_1023, and the second case, 1024, is called the representation Excess_1024.

In IEEE standards, single precision representation of a floating-point number contains three fields: the sign, the exponent (power of 2) and the mantissa. Figure 4.8 is employed to show the format. From Fig. 4.8 it is indicated that the

Fig. 4.8 Single precision representation of a floating-point number

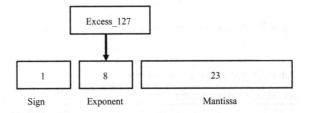

number inside the boxes is the number of bits for each field. This implies that three lengths for the sign, the exponent and the mantissa are, respectively, one bit, eight bits and twenty-three bits. The procedure, that is applied to store a normalized floating-point number by means of single precision format, contains three main steps. The first main step is to store the sign as 0 (positive) or 1 (negative). Next, the second main step is to store the exponent (power of 2) as Excess_127. The third step is to store the mantissa as an unsigned integer. From the statements above, in bio-molecular computer, the following definition is used to define the range for single precision format of a floating-point number.

Definition 4-5: Suppose that a 32-bit binary number, $x_{32} \ldots x_1$ is used to represent a floating-point number of 32 bits in form of single precision format based on Excess_127, where the value of each bit x_k is either 1 or 0 for $1 \leq k \leq 32$. A floating-point number of 32 bits in form of single precision format based on Excess_127 ranges between $-(2^{128} \times 1.11111111111111111111111)$ and $+(2^{128} \times 1.11111111111111111111111)$.

Algorithm 4.5 is used to construct the range of the value for a floating-point number of n bits in form of single precision format based on Excess_127. Tube T_0 is an empty tube, and it is regarded as one input tube of **Algorithm 4.5**. The second parameter in **Algorithm 4.5**, n, is employed to represent the number of bits for a floating-point number in form of single precision format based on Excess_127.

Algorithm 4.5: Generate-Single-Precision-Floating-Point-Numbers(T_0, n)

(0a) Append-head(T_1, x_1^1).

(0b) Append-head(T_2, x_1^0).

(0c) $T_0 = \cup(T_1, T_2)$.

(1) **For** $k = 2$ **to** n

    (1a) Amplify(T_0, T_1, T_2).

    (1b) Append-head(T_1, x_k^1).

    (1c) Append-head(T_2, x_k^0).

    (1d) $T_0 = \cup(T_1, T_2)$.

EndFor

EndAlgorithm

Consider that a floating-point number of 32 bits in form of single precision format based on Excess_127 ranges between $-(2^{128} \times 1.11111111111111111111111)$ and $+(2^{128} \times 1.11111111111111111111111)$. Tube T_0 is an empty tube and is regarded as an input tube of **Algorithm 4.5**. The value for n is 32 and is regarded as the second parameter in **Algorithm 4.5**. After the first execution for Step (0a) and Step (0b) is performed, tube $T_1 = \{x_1^1\}$ and tube $T_2 = \{x_1^0\}$. Then, after the first execution of Step (0c) is implemented, tube $T_0 = \{x_1^1, x_1^0\}$, tube $T_1 = \varnothing$ and tube $T_2 = \varnothing$.

Step (1) is the main loop and its lower bound and the upper bound are, respectively, two and thirty-two, so Steps (1a) through (1d) will be run 31 times. After the first execution of Step (1a) is finished, tube $T_0 = \varnothing$, tube $T_1 = \{x_1^1, x_1^0\}$ and tube $T_2 = \{x_1^1, x_1^0\}$. Next, after the first execution for Step (1b) and Step (1c) is performed, tube $T_1 = \{x_2^1 x_1^1, x_2^1 x_1^0\}$ and tube $T_2 = \{x_2^0 x_1^1, x_2^0 x_1^0\}$. After the first execution of Step (1d) is implemented, tube $T_0 = \{x_2^1 x_1^1, x_2^1 x_1^0, x_2^0 x_1^1, x_2^0 x_1^0\}$, tube $T_1 = \varnothing$ and tube $T_2 = \varnothing$.

Then, after the second execution of Step (1a) is performed, tube $T_0 = \varnothing$, tube $T_1 = \{x_2^1 x_1^1, x_2^1 x_1^0, x_2^0 x_1^1, x_2^0 x_1^0\}$ and tube $T_2 = \{x_2^1 x_1^1, x_2^1 x_1^0, x_2^0 x_1^1, x_2^0 x_1^0\}$. After the rest of operations are finished, tube $T_1 = \varnothing$, tube $T_2 = \varnothing$ and the result for tube T_0

is shown in Table 4.6. In Table 4.6, for this bit pattern, "$x_{32}^1 \, x_{31}^1 \, x_{30}^1 \, x_{29}^1 \, x_{28}^1 \, x_{27}^1 \, x_{26}^1$ $x_{25}^1 \, x_{24}^1 \, x_{23}^1 \, x_{22}^1 \, x_{21}^1 \, x_{20}^1 \, x_{19}^1 \, x_{18}^1 \, x_{17}^1 \, x_{16}^1 \, x_{15}^1 \, x_{14}^1 \, x_{13}^1 \, x_{12}^1 \, x_{11}^1 \, x_{10}^1 \, x_9^1 \, x_8^1 \, x_7^1 \, x_6^1 \, x_5^1 \, x_4^1 \, x_3^1 \, x_2^1 \, x_1^1$", the leftmost bit is the sign ($-$). The next 8 bits, "$x_{31}^1 \, x_{30}^1 \, x_{29}^1 \, x_{28}^1 \, x_{27}^1 \, x_{26}^1 \, x_{25}^1 \, x_{24}^1$", that subtract 127_{10} is the exponent (128_{10}). The next 23 bits are the mantissa. So, this bit pattern is applied to represent $-(2^{128} \times 1.11111111111111111111111)$. Similarly, in Table 4.6, for that bit pattern, "$x_{32}^0 \, x_{31}^1 \, x_{30}^1 \, x_{29}^1 \, x_{28}^1 \, x_{27}^1 \, x_{26}^1 \, x_{25}^1 \, x_{24}^1 \, x_{23}^1$ $x_{22}^1 \, x_{21}^1 \, x_{20}^1 \, x_{19}^1 \, x_{18}^1 \, x_{17}^1 \, x_{16}^1 \, x_{15}^1 \, x_{14}^1 \, x_{13}^1 \, x_{12}^1 \, x_{11}^1 \, x_{10}^1 \, x_9^1 \, x_8^1 \, x_7^1 \, x_6^1 \, x_5^1 \, x_4^1 \, x_3^1 \, x_2^1 \, x_1^1$", it is also applied to represent $+(2^{128} \times 1.11111111111111111111111)$. **Lemma 4-5** is applied to prove correction of **Algorithm 4.5**.

Table 4.6 The result for tube T_0 is generated by **Algorithm 4.5**

Tube	The result is generated by Algorithm 4.5
T_0	$\{x_{32}^1 \, x_{31}^1 \, x_{30}^1 \, x_{29}^1 \, x_{28}^1 \, x_{27}^1 \, x_{26}^1 \, x_{25}^1 \, x_{24}^1 \, x_{23}^1 \, x_{22}^1 \, x_{21}^1 \, x_{20}^1 \, x_{19}^1 \, x_{18}^1 \, x_{17}^1 \, x_{16}^1 \, x_{15}^1 \, x_{14}^1 \, x_{13}^1 \, x_{12}^1 \, x_{11}^1 \, x_{10}^1 \, x_9^1 \, x_8^1 \, x_7^1$ $x_6^1 \, x_5^1 \, x_4^1 \, x_3^1 \, x_2^1 \, x_1^1$ ... $x_{32}^0 \, x_{31}^1 \, x_{30}^1 \, x_{29}^1 \, x_{28}^1 \, x_{27}^1 \, x_{26}^1 \, x_{25}^1 \, x_{24}^1 \, x_{23}^1 \, x_{22}^1 \, x_{21}^1 \, x_{20}^1 \, x_{19}^1 \, x_{18}^1 \, x_{17}^1 \, x_{16}^1 \, x_{15}^1 \, x_{14}^1 \, x_{13}^1 \, x_{12}^1 \, x_{11}^1 \, x_{10}^1 \, x_9^1 \, x_8^1 \, x_7^1$ $x_6^1 \, x_5^1 \, x_4^1 \, x_3^1 \, x_2^1 \, x_1^1\}$

Lemma 4-5: **Algorithm 4.5** *can be applied to construct the range of the value for a floating-point number of n bits in form of single precision format based on Excess_127.*

Proof Refer to **Lemma 4-1**.

4.4.2 Introduction for Double Precision of Floating-Point Numbers

In IEEE standards, double precision representation of a floating-point number contains three fields: the sign, the exponent (power of 2) and the mantissa. Figure 4.9 is used to show the format. From Fig. 4.9, the number inside the boxes is the number of bits for each field. This is to say that three lengths for the sign, the exponent and the mantissa are, respectively, one bit, eleven bits and fifty-two bits. The procedure, that is used to store a normalized floating-point number by means of double precision format, contains three steps. The first step is to store the sign as 0 (positive) or 1 (negative). Next, the second step is to store the exponent (power of 2) as Excess_1023. The third step is to store the mantissa as an unsigned integer. From the statements above, in bio-molecular computer, the following definition is applied to define the range for double precision format of a floating-point number.

Fig. 4.9 Double precision representation of a floating-point number

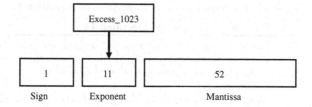

Definition 4-6: Suppose that a 64-bit binary number, $x_{64} \ldots x_1$ is applied to represent a floating-point number of 64 bits in form of double precision format based on Excess_1023, where the value of each bit x_k is either 1 or 0 for $1 \leq k \leq 64$. A floating-point number of 64 bits in form of double precision format based on Excess_1023 ranges between $-(2^{1024} \times 1.1111111111111111111111111111111111111111111111111$ $111111111)$ and $+(2^{1024} \times 1.1111111111111111111111111111111111111111111111111$ $11111111)$.

Algorithm 4.6 is employed to construct the range of the value for a floating-point number of n bits in form of double precision format based on Excess_1023. Tube T_0 is an empty tube, and it is regarded as one input tube of **Algorithm 4.6**. The second parameter in **Algorithm 4.6**, n, is employed to represent the number of bits for a floating-point number in form of double precision format based on Excess_1023.

Algorithm 4.6: **Generate-Double-Precision-Floating-Point-Numbers(T_0, n)**
(0a) Append-head($T_1, x_1{}^1$).
(0b) Append-head($T_2, x_1{}^0$).
(0c) $T_0 = \cup(T_1, T_2)$.
(1) **For** $k = 2$ **to** n
    (1a) Amplify(T_0, T_1, T_2).
    (1b) Append-head($T_1, x_k{}^1$).
    (1c) Append-head($T_2, x_k{}^0$).
    (1d) $T_0 = \cup(T_1, T_2)$.
EndFor
EndAlgorithm

Consider that a floating-point number of 64 bits in form of double precision format based on Excess_1023 ranges between $-(2^{1024} \times 1.1111111111111111$ $111111111111111111111111111111111111111)$ and $+(2^{1024} \times 1.11111111111111$ $1111111111111111111111111111111111111111111)$. Tube T_0 is an empty tube and is regarded as an input tube of **Algorithm 4.6**. The value for n is 64 and is regarded as the second parameter in **Algorithm 4.6**. After the first execution for Step (0a) and Step (0b) is finished, tube $T_1 = \{x_1^1\}$ and tube $T_2 = \{x_1^0\}$. Next, after the first execution of Step (0c) is implemented, tube $T_0 = \{x_1^1, x_1^0\}$, tube $T_1 = \varnothing$ and tube $T_2 = \varnothing$.

Step (1) is the main loop and its lower and upper bounds are, respectively, two and sixty-four, so Steps (1a) through (1d) will be run 63 times. After the first

Table 4.7 The result for tube T_0 is generated by **Algorithm 4.6**

Tube	The result is generated by **Algorithm 4.6**
T_0	$\{x_{64}^1 \, x_{63}^1 \, x_{62}^1 \, x_{61}^1 \, x_{60}^1 \, x_{59}^1 \, x_{58}^1 \, x_{57}^1 \, x_{56}^1 \, x_{55}^1 \, x_{54}^1 \, x_{53}^1 \, x_{52}^1 \, x_{51}^1 \, x_{50}^1 \, x_{49}^1 \, x_{48}^1 \, x_{47}^1 \, x_{46}^1 \, x_{45}^1 \, x_{44}^1 \, x_{43}^1 \, x_{42}^1 \, x_{41}^1 \, x_{40}^1$ $x_{39}^1 \, x_{38}^1 \, x_{37}^1 \, x_{36}^1 \, x_{35}^1 \, x_{34}^1 \, x_{33}^1 \, x_{32}^1 \, x_{31}^1 \, x_{30}^1 \, x_{29}^1 \, x_{28}^1 \, x_{27}^1 \, x_{26}^1 \, x_{25}^1 \, x_{24}^1 \, x_{23}^1 \, x_{22}^1 \, x_{21}^1 \, x_{20}^1 \, x_{19}^1 \, x_{18}^1 \, x_{17}^1 \, x_{16}^1$ $x_{15}^1 \, x_{14}^1 \, x_{13}^1 \, x_{12}^1 \, x_{11}^1 \, x_{10}^1 \, x_9^1 \, x_8^1 \, x_7^1 \, x_6^1 \, x_5^1 \, x_4^1 \, x_3^1 \, x_2^1 \, x_1^1$ ... $x_{64}^0 \, x_{63}^1 \, x_{62}^1 \, x_{61}^1 \, x_{60}^1 \, x_{59}^1 \, x_{58}^1 \, x_{57}^1 \, x_{56}^1 \, x_{55}^1 \, x_{54}^1 \, x_{53}^1 \, x_{52}^1 \, x_{51}^1 \, x_{50}^1 \, x_{49}^1 \, x_{48}^1 \, x_{47}^1 \, x_{46}^1 \, x_{45}^1 \, x_{44}^1 \, x_{43}^1 \, x_{42}^1 \, x_{41}^1 \, x_{40}^1$ $x_{39}^1 \, x_{38}^1 \, x_{37}^1 \, x_{36}^1 \, x_{35}^1 \, x_{34}^1 \, x_{33}^1 \, x_{32}^1 \, x_{31}^1 \, x_{30}^1 \, x_{29}^1 \, x_{28}^1 \, x_{27}^1 \, x_{26}^1 \, x_{25}^1 \, x_{24}^1 \, x_{23}^1 \, x_{22}^1 \, x_{21}^1 \, x_{20}^1 \, x_{19}^1 \, x_{18}^1 \, x_{17}^1 \, x_{16}^1$ $x_{15}^1 \, x_{14}^1 \, x_{13}^1 \, x_{12}^1 \, x_{11}^1 \, x_{10}^1 \, x_9^1 \, x_8^1 \, x_7^1 \, x_6^1 \, x_5^1 \, x_4^1 \, x_3^1 \, x_2^1 \, x_1^1\}$

execution of Step (1a) is implemented, tube $T_0 = \varnothing$, tube $T_1 = \{x_1^1, x_1^0\}$ and tube $T_2 = \{x_1^1, x_1^0\}$. Next, after the first execution for Step (1b) and Step (1c) is finished, tube $T_1 = \{x_2^1 \, x_1^1, x_2^1 \, x_1^0\}$ and tube $T_2 = \{x_2^0 \, x_1^1, x_2^0 \, x_1^0\}$. After the first execution of Step (1d) is implemented, tube $T_0 = \{x_2^1 \, x_1^1, x_2^1 \, x_1^0, x_2^0 \, x_1^1, x_2^0 \, x_1^0\}$, tube $T_1 = \varnothing$ and tube $T_2 = \varnothing$. Then, after the second execution of Step (1a) is performed, tube $T_0 = \varnothing$, tube $T_1 = \{x_2^1 \, x_1^1, x_2^1 \, x_1^0, x_2^0 \, x_1^1, x_2^0 \, x_1^0\}$ and tube $T_2 = \{x_2^1 \, x_1^1, x_2^1 \, x_1^0, x_2^0 \, x_1^1, x_2^0 \, x_1^0\}$. After the rest of operations are performed, tube $T_1 = \varnothing$, tube $T_2 = \varnothing$ and the result for tube T_0 is shown in Table 4.7.

In Table 4.7, for this bit pattern, "$x_{64}^1 \, x_{63}^1 \, x_{62}^1 \, x_{61}^1 \, x_{60}^1 \, x_{59}^1 \, x_{58}^1 \, x_{57}^1 \, x_{56}^1 \, x_{55}^1 \, x_{54}^1 \, x_{53}^1 \, x_{52}^1 \, x_{51}^1 \, x_{50}^1 \, x_{49}^1 \, x_{48}^1 \, x_{47}^1 \, x_{46}^1 \, x_{45}^1 \, x_{44}^1 \, x_{43}^1 \, x_{42}^1 \, x_{41}^1 \, x_{40}^1 \, x_{39}^1 \, x_{38}^1 \, x_{37}^1 \, x_{36}^1 \, x_{35}^1 \, x_{34}^1 \, x_{33}^1 \, x_{32}^1 \, x_{31}^1 \, x_{30}^1 \, x_{29}^1 \, x_{28}^1 \, x_{27}^1 \, x_{26}^1 \, x_{25}^1 \, x_{24}^1 \, x_{23}^1 \, x_{22}^1 \, x_{21}^1 \, x_{20}^1 \, x_{19}^1 \, x_{18}^1 \, x_{17}^1 \, x_{16}^1 \, x_{15}^1 \, x_{14}^1 \, x_{13}^1 \, x_{12}^1 \, x_{11}^1 \, x_{10}^1 \, x_9^1 \, x_8^1 \, x_7^1 \, x_6^1 \, x_5^1 \, x_4^1 \, x_3^1 \, x_2^1 \, x_1^1$", the leftmost bit is the sign ($-$). The next 11 bits that subtract 1023_{10} is the exponent (1024_{10}). The next 52 bits are the mantissa. So, this bit pattern is applied to represent $-(2^{1024} \times 1.1111111111111111111111111111111111111111111111111111)$. Similarly, in Table 4.7, for that bit pattern, "$x_{64}^1 \, x_{63}^1 \, x_{62}^1 \, x_{61}^1 \, x_{60}^1 \, x_{59}^1 \, x_{58}^1 \, x_{57}^1 \, x_{56}^1 \, x_{55}^1 \, x_{54}^1 \, x_{53}^1 \, x_{52}^1 \, x_{51}^1 \, x_{50}^1 \, x_{49}^1 \, x_{48}^1 \, x_{47}^1 \, x_{46}^1 \, x_{45}^1 \, x_{44}^1 \, x_{43}^1 \, x_{42}^1 \, x_{41}^1 \, x_{40}^1 \, x_{39}^1 \, x_{38}^1 \, x_{37}^1 \, x_{36}^1 \, x_{35}^1 \, x_{34}^1 \, x_{33}^1 \, x_{32}^1 \, x_{31}^1 \, x_{30}^1 \, x_{29}^1 \, x_{28}^1 \, x_{27}^1 \, x_{26}^1 \, x_{25}^1 \, x_{24}^1 \, x_{23}^1 \, x_{22}^1 \, x_{21}^1 \, x_{20}^1 \, x_{19}^1 \, x_{18}^1 \, x_{17}^1 \, x_{16}^1 \, x_{15}^1 \, x_{14}^1 \, x_{13}^1 \, x_{12}^1 \, x_{11}^1 \, x_{10}^1 \, x_9^1 \, x_8^1 \, x_7^1 \, x_6^1 \, x_5^1 \, x_4^1 \, x_3^1 \, x_2^1 \, x_1^1$", it is also used to represent $+(2^{1024} \times 1.1111111111111111111111111111111111111111111111111111)$.
Lemma 4-6 is applied to prove correction of **Algorithm 4.6**.

Lemma 4-6: **Algorithm 4.6** *can be applied to construct the range of the value for a floating-point number of n bits in form of double precision format based on Excess_1023.*

Proof Refer to **Lemma 4-1**.

4.5 Summary

In this chapter an introduction to the decimal system and the binary system was provided, and a description for how numbers between the decimal system and the binary system were converted and how numbers in the form of bit patterns were stored inside a tube in bio-molecular computer was also provided. We first introduced the decimal system and the binary number. We then described conversion of numbers between the decimal system and the binary number. We also introduced unsigned integers, sign-and-magnitude integers, one's complement integers and two's complement integers. We then described **Algorithm 4.1** and its proof to explain how the range of the value for an unsigned integer of n bits was constructed. We also introduced **Algorithm 4.2** and its proof to show how the range of the value for a sign-and-magnitude integer of n bits was constructed. We then described **Algorithm 4.3** and its proof to reveal how the range of the value for a one's complement integer of n bits was constructed. We also introduced **Algorithm 4.4** and its proof to demonstrate how the range of the value for a two's complement integer of n bits was constructed. We then described representation of floating-point numbers based on that the Institute of Electrical and Electronics Engineers (IEEE) had defined single precision and double precision for storing floating-point numbers. We also introduced **Algorithm 4.5** and its proof to explain how the range of the value for a floating-point number of n bits in form of single precision format based on Excess_127 was constructed. We then described **Algorithm 4.6** and its proof to show how the range of the value for a floating-point number of n bits in form of double precision format based on Excess_1023 was constructed.

4.6 Bibliographical Notes

In this chapter, for a more detailed description to unsigned integers, sign-and-magnitude integers, one's complement integers and two's complement integers, the recommended books are Brown and Vranesic (2007), Forouzan and Mosharraf (2008), Mano (1979), Mano (1993), Null and Lobur (2010), Reed (2008), Shiva (2008), Stalling (2000). The books which were written by the authors in Brown and Vranesic (2007), Forouzan and Mosharraf (2008), Mano (1979), Mano (1993), Null and Lobur (2010), Reed (2008), Shiva (2008), Stalling (2000) are good introduction to representation of floating-point numbers.

4.7 Exercises

4.1 Concisely explain what a decimal system and a binary system are.

4.2 Convert two decimal numbers 128 and 33 to their corresponding binary numbers with *eight* bits.

4.3 Convert two binary numbers 10000000 and 00100001 to their corresponding decimal numbers.

4.4 Write a digital program to convert a decimal number to a binary number.

4.5 Write a digital program to convert a binary number to a decimal number.

4.6 It is assumed that an n-bit binary number, $x_n \ldots x_1$ is used to represent an unsigned integer of n bits, where the value of each bit x_k is either 1 or 0 for $1 \leq k \leq n$. The bits x_n and x_1 is applied to represent, respectively, the most significant bit and the least significant bit for an unsigned integer of n bits. An unsigned integer of n bits ranges between 0 and $2^n - 1$. Write a bio-molecular program to construct the range of the value for an unsigned integer of *three* bits.

4.7 It is assumed that an n-bit binary number, $x_n \ldots x_1$ is used to represent a sign-and-magnitude integer of n bits, where the value of each bit x_k is either 1 or 0 for $1 \leq k \leq n$. The bit x_n is used to represent the sign, and the bits x_{n-1} and x_1 is applied to represent, respectively, the most significant bit and the least significant bit for a sign-and-magnitude integer of n bits. A sign-and-magnitude integer of n bits ranges between $-(2^{n-1} - 1)$ and $+(2^{n-1} - 1)$. Write a bio-molecular program to construct the range of the value for a sign-and-magnitude integer of *three* bits.

4.8 It is supposed that an n-bit binary number, $x_n \ldots x_1$ is used to represent a one's complement integer of n bits, where the value of each bit x_k is either 1 or 0 for $1 \leq k \leq n$. A one's complement integer of n bits ranges between $-(2^{n-1} - 1)$ and $+(2^{n-1} - 1)$ Write a bio-molecular program to construct the range of the value for a one's complement integer of *three* bits.

4.9 It is assumed that an n-bit binary number, $x_n \ldots x_1$ is used to represent a two's complement integer of n bits, where the value of each bit x_k is either 1 or 0 for $1 \leq k \leq n$. A two's complement integer of n bits ranges between $-(2^{n-1} - 1)$ and $+(2^{n-1} - 1)$. Write a bio-molecular program to construct the range of the value for a two's complement integer of *three* bits.

4.10 It is supposed that a 32-bit binary number, $x_{32} \ldots x_1$ is used to represent a floating-point number of 32 bits in form of single precision format based on Excess_127, where the value of each bit x_k is either 1 or 0 for $1 \leq k \leq 32$. A floating-point number of 32 bits in form of single precision format based on Excess_127 ranges between $-(2^{128} \times 1.11111111111111111111111)$ and $+(2^{128} \times 1.11111111111111111111111)$. Write a bio-molecular program to construct the range of the value for a floating-point number of 32 bits in form of single precision format based on Excess_127.

4.11 It is assumed that a 64-bit binary number, $x_{64} \ldots x_1$ is applied to represent a floating-point number of 64 bits in form of double precision format based on Excess_1023, where the value of each bit x_k is either 1 or 0 for $1 \leq k \leq 64$. A floating-point number of 64 bits in form of double precision format based on Excess_1023 ranges between $-(2^{1024} \times 1.1111111111111111111111$ $1111111111111111111111111111111111)$ and $+(2^{1024} \times 1.1111111111111111$ $1111111111111111111111111111111111111111)$. Write a bio-molecular program to construct the range of the value for a floating-point number of 64 bits in form of double precision format based on Excess_1023.

References

B. Forouzan, F. Mosharraf, *Foundations of Computer Science*, 2nd edn. (Thomson, London, 2008). ISBN: 978-1-84480-700-0

D. Reed, *A Balanced Introduction to Computer Science*. (Pearson Prentice Hall, New Jersy, 2008). ISBN: 9780136017226

L. Null, J. Lobur: *Essentials of Computer Organization and Architecture*. (Jones and Bartlett Learning, Sudbury, 2010). ISBN: 978-1449600068

M.M. Mano, *Digital Logic and Computer Design*. (Prentice-Hall, New Jersy, 1979). ISBN: 0-13-214510-3

M.M. Mano, *Computer System Architecture*. (Prentice Hall, New Jersy, 1993). ISBN: 978-0131755635

S. Brown, Z. Vranesic, *Fundamentals of Digital Logic with Verilog Design*. (McGraw-Hill, New York, 2007). ISBN: 978-0077211646

S.G. Shiva, *Computer Organization, Design, and Architecture*. (CRC Press, Boca Raton, 2008). ISBN: 9780849304163

W. Stalling, *Computer Organization and Architecture*. (Prentice Hall, New Jersy, 2000). ISBN: 978-0132936330

Chapter 5
Introduction to Arithmetic Operations on Bits on Bio-molecular Computer

In previous chapters, we showed how to represent and to store different types of data in a bio-molecular computer. In this chapter and Chap. 6, we show how to operate on bits in a bio-molecular computer. Operations on bits in a bio-molecular computer can be classified as two main computations: arithmetic operations and logical operations. Figure 5.1 is applied to explain two broad categories for operations of bits.

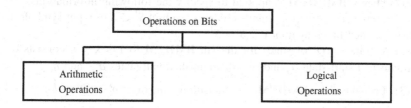

Fig. 5.1 Operations on bits

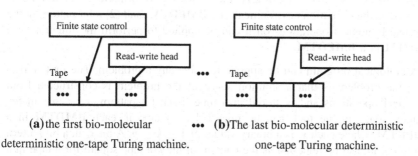

    (a) the first bio-molecular ••• **(b)**The last bio-molecular deterministic
deterministic one-tape Turing machine.   one-tape Turing machine.

Fig. 5.2 Schematic representation of a *bio-molecular parallel deterministic one-tape Turing machine* (abbreviated **BMPDTM**)

W.-L. Chang and A. V. Vasilakos, *Molecular Computing*, Studies in Big Data 4,   53
DOI: 10.1007/978-3-319-05122-2_5, © Springer International Publishing Switzerland 2014

In order to clearly explain how operations on bits perform arithmetic computations and logical computations, we will need to fix a particular model for arithmetic computations and logical computations. The model we choose is the *bio-molecular parallel deterministic one-tape Turing machine* (abbreviated **BMPDTM**), which is pictured schematically in Fig. 5.2. This machine includes n bio-molecular *deterministic one-tape Turing machines*. Each bio-molecular *deterministic one-tape Turing machine* (abbreviated **BMDTM**) contains a finite state control, a read-write head, and a tape made up of a two-way infinite sequence of tape squares (shown in Fig. 5.2).

Each bit pattern encoded in a tube in a bio-molecular computer can be regarded as a tape in a bio-molecular deterministic one-tape Turing machine. For each tape in a **BMPDTM**, each biological operation can be applied to simultaneously perform the function of the corresponding finite state control and read-write head. This is to say that n bio-molecular deterministic one-tape Turing machines simultaneously are run in parallel model. **Definition 5.1** is used to denote the function for each **BMDTM** in a **BMPDTM**, and **Definition 5.2** is applied to describe the execution environment of a **BMPDTM**.

Definition 5.1 A biological program made up biological operations for each **BMDTM** in a **BMPDTM** is applied to specify the following information.

(1) A finite set S of tape symbols contains a subset $S^1 \subset S$ of input symbols and a distinguished blank symbol $\theta \in S - S^1$.

(2) A finite set Q^k of states for the kth **BMDTM** to $1 \le k \le n$ consists of a distinguished start-state q_0^k and two distinguished halt-states q_{yes}^k and q_{no}^k.

(3) For the kth **BMDTM**, a transition function $\delta^k : (Q^k - \left\{ q_{yes}^k, q_{no}^k \right\})$ $\times S^1 \to Q^k \times S^1 \times \{-1, +1\}$.

Definition 5.2 A biological program made up biological operations for a **BMPDTM** is employed to specify the following information: (1) a tube is regarded as the execution environment of a **BMPDTM**, and (2) a bit pattern in the tube and biological operations on it can be applied to perform the function of a **BMDTM** in a **BMPDTM**.

A biological program can be applied to solve any a problem. The first step to solve the problem is that a solution space of the problem is constructed from biological operations and is stored in a tube. Each bit pattern, $x_n \ldots x_1$, in the solution space in the tube can be regarded as a tape in each **BMDTM** in a **BMPDTM**, where each x_k is a binary value for $1 \le k \le n$. A bit x_k in a bit pattern in a tube can also be regarded as the content of the kth tape square for a tape in a **BMDTM** in a **BMPDTM**. This is to say that each input, $x_n \ldots x_1$, for each **BMDTM** in a **BMPDTM** is placed in the corresponding tape squares 1 through n.

The biological program starts to simultaneously run its biological operation in the start-state q_0^k of each **BMDTM** in a **BMPDTM**, with the read-write head of each **BMDTM** in a **BMPDTM** scanning the content of the corresponding tape. Then, all of the operations for the biological program are run in a step-by-step manner in each **BMDTM** in a **BMPDTM**. Each operation in the biological program is simultaneously run in each **BMDTM** in a **BMPDTM**. If for the kth **BMDTM** in a **BMPDTM** the current state $q_j^k \in Q^k$ and is either q_{yes}^k or q_{no}^k, then the operation has ended, with the answer that is "yes" if $q_j^k = q_{yes}^k$ and "no" if $q_j^k = q_{no}^k$. Otherwise for the kth **BMDTM** in a **BMPDTM** the current state $q_j^k \in Q^k - \left\{ q_{yes}^k, q_{no}^k \right\}$, some tape symbol $s \in S$ in the tape square being scanned, and the value of $d^k \left(q_j^k, s \right)$ is denoted as "(q_a^k, s^1, Δ)". If $\Delta = 1$, the read-write head writes s^1 in the *right new* tape square and the position of the read-write head is moved the right new tape square. If $\Delta = -1$, the read-write head writes s^1 in the *left new* tape square and the position of the read-write head is moved the left new tape square. At the same time, the finite state control changes its state from q_j^k to q_a^k. This implies that one step of the computation for the operation is completed. If there are still new operations in the biological program, then the next operation will be continued to proceed in each **BMDTM** in a **BMPDTM**.

5.1 Introduction to Arithmetic Operations on Bio-molecular Computer

Arithmetic operations contain adding, subtracting, multiplying, dividing, and so on. These operations can be employed to deal with integers and floating-point numbers. However, we mainly focus only on addition and subtraction on integers. An extended discussion of multiplication and division is beyond the scope of this book, and if a reader is interested in an extended discussion of multiplication and division, then please refer to (Chang et al. 2005; Ho 2005). The following sections will be used to describe how to perform addition and subtraction on integers.

5.2 Introduction to Addition on Unsigned Integers on Bio-molecular Computer

A one-bit adder is a Boolean function that forms the arithmetic sum of three input bits. It includes three inputs and two outputs. Two of the input bits represent augend and addend bits to be added, respectively. The third input represents the carry from the previous lower significant position. The first output gives the value of the sum for augend and addend bits to be added. The second output gives the value of the carry to augend and addend bits to be added. The truth table of the one-bit adder is as follows:

Table 5.1 The truth table of a one-bit adder

Augend bit	Addend bit	Previous carry bit	Sum bit	Carry bit
0	0	0	0	0
0	0	1	1	0
0	1	0	1	0
0	1	1	0	1
1	0	0	1	0
1	0	1	0	1
1	1	0	0	1
1	1	1	1	1

The one-bit adder figures out the sum and the carry of two bits and a previous carry. The first operand of n-bits and the second operand of n-bits each can be added by means of performing this one-bit adder of n times. This is to say that a parallel adder of n-bits is also a Boolean function that performs the arithmetic sum for the first operand of n-bits and the second operand of n-bits. The following subsections will be applied to describe how to finish a one-bit adder and a parallel adder of n-bits.

5.2.1 The Construction of a Parallel One-bit Adder on Unsigned Integers on Bio-molecular Computer

Suppose that two one-bit binary numbers, x_k and y_k, are used to represent the first input of a one-bit adder, and the first output of a one-bit adder for $1 \leq k \leq n$, respectively, a one-bit binary number, p_k, is applied to represent the second input of a one-bit adder, and two one-bit binary numbers, z_k and z_{k-1}, are employed to represent the second output and the third input of a one-bit adder, respectively. For the sake of convenience, assume that z_k^1 contains the value of z_k to be 1 and z_k^0 contains the value of z_k to be 0. Similarly, also suppose that z_{k-1}^1 contains the value of z_{k-1} to be 1 and z_{k-1}^0 contains the value of z_{k-1} to be 0. Assume that p_k^1 denotes the value of p_k to be 1 and $p_k{}^0$ defines the value of p_k to be 0. Similarly, also suppose that y_k^1 denotes the value of y_k to be 1 and $y_k{}^0$ defines the value of y_k to be 0. Assume that x_k^1 denotes the value of x_k to be 1 and $x_k{}^0$ defines the value of x_k to be 0. The following algorithm is proposed to perform the Boolean function of a parallel one-bit adder.

Algorithm 5.1: ParallelOneBitAdder(T_0, k)

(1) $T_1 = +(T_0, x_k^1)$ and $T_2 = -(T_0, x_k^1)$.

(2) $T_3 = +(T_1, p_k^1)$ and $T_4 = -(T_1, p_k^1)$.

(3) $T_5 = +(T_2, p_k^1)$ and $T_6 = -(T_2, p_k^1)$.

(4) $T_7 = +(T_3, z_{k-1}^1)$ and $T_8 = -(T_3, z_{k-1}^1)$.

(5) $T_9 = +(T_4, z_{k-1}^1)$ and $T_{10} = -(T_4, z_{k-1}^1)$.

(6) $T_{11} = +(T_5, z_{k-1}^1)$ and $T_{12} = -(T_5, z_{k-1}^1)$.

(7) $T_{13} = +(T_6, z_{k-1}^1)$ and $T_{14} = -(T_6, z_{k-1}^1)$.

(8a) **If** (Detect(T_7) = = "yes") **then**

    (8) Append-head(T_7, y_k^1) and Append-head(T_7, z_k^1).

EndIf

(9a) **If** (Detect(T_8) = = "yes") **then**

    (9) Append-head(T_8, y_k^0) and Append-head(T_8, z_k^1).

EndIf

(10a) **If** (Detect(T_9) = = "yes") **then**

    (10) Append-head(T_9, y_k^0) and Append-head(T_9, z_k^1).

EndIf

(11a) **If** (Detect(T_{10}) = = "yes") **then**

    (11) Append-head(T_{10}, y_k^1) and Append-head(T_{10}, z_k^0).

EndIf

(12a) **If** (Detect(T_{11}) = = "yes") **then**

    (12) Append-head(T_{11}, y_k^0) and Append-head(T_{11}, z_k^1).

EndIf

(13a) **If** (Detect(T_{12}) = = "yes") **then**

    (13) Append-head(T_{12}, y_k^1) and Append-head(T_{12}, z_k^0).

EndIf

(14a) **If** (Detect(T_{13}) = = "yes") **then**

    (14) Append-head(T_{13}, y_k^1) and Append-head(T_{13}, z_k^0).

EndIf

(15a) **If** (Detect(T_{14}) = = "yes") **then**

    (15) Append-head(T_{14}, y_k^0) and Append-head(T_{14}, z_k^0).

EndIf

(16) $T_0 = \cup(T_7, T_8, T_9, T_{10}, T_{11}, T_{12}, T_{13}, T_{14})$.

EndAlgorithm

Lemma 5-1: *The algorithm,* **ParallelOneBitAdder**(T_0, k), *can be used to finish the Boolean function of a parallel one-bit adder.*

Proof The algorithm **ParallelOneBitAdder**(T_0, k) is implemented via the *extract, append-head, detect* and *merge* operations. Steps (1) through (7) employ the *extract* operations to form some different test tubes including different inputs (T_1 to T_{14}). That is, T_1 includes all of the inputs that have $x_k = 1$, T_2 contains all of the inputs that have $x_k = 0$, T_3 consists of those that have $x_k = 1$ and $p_k = 1$, T_4 includes those that have $x_k = 1$ and $p_k = 0$, T_5 contains those that have $x_k = 0$ and $p_k = 1$, T_6 consists of those that have $x_k = 0$ and $p_k = 0$, T_7 includes those that have $x_k = 1$, $p_k = 1$ and $z_{k-1} = 1$, T_8 contains those that have $x_k = 1$, $p_k = 1$ and $z_{k-1} = 0$, T_9 consists of those that have $x_k = 1$, $p_k = 0$ and $z_{k-1} = 1$, T_{10} consists of those that have $x_k = 1$, $p_k = 0$ and $z_{k-1} = 0$, T_{11} includes those that have $y_k = 0$, $p_k = 1$ and $z_{k-1} = 1$, T_{12} contains those that have $x_k = 0$, $p_k = 1$ and $z_{k-1} = 0$, T_{13} consists of those that have $x_k = 0$, $p_k = 0$ and $z_{k-1} = 1$, and finally, T_{14} consists of those that have $x_k = 0$, $p_k = 0$ and $z_{k-1} = 0$. Having performed Steps (1) through (7), this implies that eight different inputs of a one-bit adder as shown in Table 5.1 were poured into tubes T_7 through T_{14}, respectively.

Steps (8a) (9a) (10a) (11a) (12a) (13a) (14a) and (15a) are, respectively, used to check whether contains any input for tubes T_7, T_8, T_9, T_{10}, T_{11}, T_{12}, T_{13} and T_{14} or not. If any a "yes" is returned for those steps, then the corresponding *append-head* operations will be run. Next, Steps (8) through (15) use the *append-head* operations to append y_k^1 or y_k^0, and z_k^1 or z_k^0 onto the head of every input in the corresponding test tubes. After performing Steps (8) through (15), we can say that eight different outputs of a one-bit adder in Table 5.1 are appended into tubes T_7 through T_{14}. Finally, the execution of Step (16) applies the *merge* operation to pour tubes T_7 through T_{14} into tube T_0. Tube T_0 contains the result performing the addition of a bit. ∎

5.2.2 The Construction for a Parallel Adder of N Bits on Unsigned Integers on Bio-molecular Computer

The parallel one-bit adder introduced in Sect. 5.2.1 is used to compute the sum and the carry of two bits and a previous carry. It directly uses the truth table (Table 5.1 in Sect. 5.2.1) to perform addition of a bit. The main difference between one-bit adder on bio-molecular computer and a full adder of one-bit on Turing's machines is that the Karnaugh map and basic logic gates on Turing's machines are not applied to implement addition of a bit on bio-molecular computer. Similarly, a parallel adder of n-bits is also directly to perform the arithmetic sum for the first operand of n-bits and the second operand of n-bits by means of performing this one-bit adder of n times. The following algorithm is proposed to perform the arithmetic sum for a parallel adder of n-bits.

Algorithm 5.2: **BinaryParallelAdder**(T_0, n)

(0a) Append-head(T_1, $x_1{}^1$).

(0b) Append-head(T_2, $x_1{}^0$).

(0c) $T_0 = \cup(T_1, T_2)$.

(1) **For** $k = 2$ **to** n

    (1a) Amplify(T_0, T_1, T_2).

    (1b) Append-head(T_1, $x_k{}^1$).

    (1c) Append-head(T_2, $x_k{}^0$).

    (1d) $T_0 = \cup(T_1, T_2)$.

EndFor

(2) **For** $k = 1$ **to** n

    (2a) Append-head(T_0, p_k).

EndFor

(2b) Append-head(T_0, $z_0{}^0$).

(3) **For** $k = 1$ **to** n

    (3a) **ParallelOneBitAdder**(T_0, k).

EndFor

EndAlgorithm

Lemma 5-2: *The algorithm,* **BinaryParallelAdder**(T_0, n), *can be applied to perform the Boolean function to a binary parallel adder of n bits.*

Proof Steps (0a) through (1d) are used to construct 2^n combinations of n bits. After they are performed, tube T_0 includes those inputs encoding 2^n unsigned integers of n bits (the range of values for them is from 0 to $2^n - 1$). Step (2) is the first loop and is used to construct addend bits for an adder of n bits. On each execution of Step (2a), it applies the "Append-head" operation to append the value "0" or "1" for the kth bit of an addend into the head of each bit pattern in tube T_0.

When the operation for an adder of n bits, the least significant position for the augend and the addend is added, the previous carry bit must be 0. On the execution of Step (2b), it uses the "Append-head" operation to append the value 0 of the previous carry bit, z_0, onto the head of every bit pattern in T_0. Next, Step (3) is the main loop and is mainly used to perform the Boolean function of a binary parallel adder of n bits. Each execution of Step (3a) calls the procedure, **ParallelOneBitAdder**(T_0, k), in Sect. 5.2.1 to figure out the arithmetic sum of one bit for the augend and the addend. Repeat execution of Step (3a) until the most significant bit for the augend and the addend is processed. Tube T_0 contains the result performing the arithmetic sum for the first operand of n-bits and the second operand of n-bits. ∎

5.2.3 The Power for a Parallel Adder of N Bits on Unsigned Integers on Bio-molecular Computer

Consider that eight values for an unsigned integer of three bits are, respectively, $000(0_{10})$ $(x_3^0 \, x_2^0 \, x_1^0)$, $001(1_{10})$ $(x_3^0 \, x_2^0 \, x_1^1)$, $010(2_{10})$ $(x_3^0 \, x_2^1 \, x_1^0)$, $011(3_{10})$ $(x_3^0 \, x_2^1 \, x_1^1)$, $100(4_{10})$ $(x_3^1 \, x_2^0 \, x_1^0)$, $101(5_{10})$ $(x_3^1 \, x_2^0 \, x_1^1)$, $110(6_{10})$ $(x_3^1 \, x_2^1 \, x_1^0)$ and $111(7_{10})$ $(x_3^1 \, x_2^1 \, x_1^1)$. We want to simultaneously add $001(1_{10})$ $(p_3^0 \, p_2^0 \, p_1^1)$ to those eight values. **Algorithm 5.2**, **BinaryParallelAdder** (T_0, n), can be applied to perform the task. Tube T_0 is an empty tube and is regarded as an input tube of **Algorithm 5.2**. The value for the second parameter, n, is three because the number of bits for representing those eight values is three. From **Definition 5.2**, the input tube T_0 can be regarded as the execution environment of a **BMPDTM**. Steps (0a) through (1d) are applied to construct a **BMPDTM** with eight bio-molecular deterministic one-tape Turing machines.

First, after the first execution for Step (0a) and Step (0b) is implemented, tube $T_1 = \{x_1^1\}$ and tube $T_2 = \{x_1^0\}$. This is to say that a **BMDTM** in the second **BMPDTM** (tube T_1) and in the third **BMPDTM** (tube T_2) is constructed. Figure 5.3 is employed to show the current status of the execution environment to the second **BMPDTM** and the third **BMPDTM**. From Fig. 5.3, the content of the first tape square for the tape in the first **BMDTM** in the second **BMPDTM** is written by its corresponding read-write head and is 1 ($x_1 = 1$), and the content of the first tape square for the tape in the first **BMDTM** in the third **BMPDTM** is written by its corresponding read-write head and is 0 ($x_1 = 0$). Simultaneously, the position of each read-write head is moved to the left new tape square and the status of each finite state control is, respectively, changed as "$x_1 = 1$" and "$x_1 = 0$".

Next, after the first execution for Step (0c) is performed, tube $T_0 = \{x_1^1, x_1^0\}$, tube $T_1 = \varnothing$ and tube $T_2 = \varnothing$. This implies that the execution environment for the first **BMDTM** in the second **BMPDTM** and the first **BMDTM** in the third **BMPDTM** becomes the first **BMPDTM**. Figure 5.4 is applied to illustrate the current status of the execution environment to the first **BMPDTM**. From Fig. 5.4, the contents to the two tapes in the execution environment of the first **BMPDTM** are not changed. Simultaneously, the position and the status for each read-write head are also not changed.

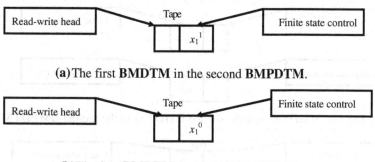

(a) The first **BMDTM** in the second **BMPDTM**.

(b) The first **BMDTM** in the third **BMPDTM**.

Fig. 5.3 Schematic representation of the current status of the execution environment to the second **BMPDTM** and the third **BMPDTM**

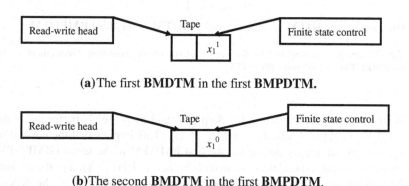

(a) The first **BMDTM** in the first **BMPDTM**.

(b) The second **BMDTM** in the first **BMPDTM**.

Fig. 5.4 Schematic representation of the current status of the execution environment to the first **BMPDTM**

Because Step (1) is the main loop and its lower and upper bounds are, respectively, two and three. Therefore, Steps (1a) through (1d) will be run two times. Then, after the first execution of Step (1a) is finished, tube $T_0 = \varnothing$, tube $T_1 = \{x_1^1, x_1^0\}$ and tube $T_2 = \{x_1^1, x_1^0\}$. This is to say that the first **BMDTM** and the second **BMDTM** in the execution environment of the first **BMPDTM** are both copied into the second **BMPDTM** and the third **BMPDTM**. Figure 5.5 is applied to show the current status of the execution environment to the second **BMPDTM** and the third **BMPDTM**. From Fig. 5.5, the contents of the first tape square for the corresponding tape of the first **BMDTM** and the corresponding tape of the second **BMDTM** in the execution environment of the second **BMPDTM** are, respectively, 1 ($x_1 = 1$) and 0 ($x_1 = 0$). The contents of the first tape square for the corresponding tape of the first **BMDTM** and the corresponding tape of the second **BMDTM** in the execution environment of the third **BMPDTM** are also, respectively, 0 ($x_1 = 0$) and 1 ($x_1 = 1$).

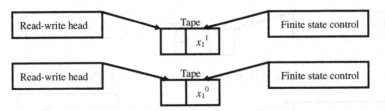

(a) The first **BMDTM** and the second **BMDTM** in the second **BMPDTM**.

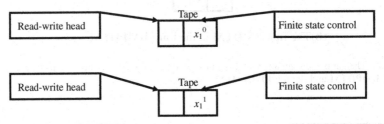

(b) The first **BMDTM** and the second **BMDTM** in the third **BMPDTM**.

Fig. 5.5 Schematic representation of the current status of the execution environment to the second **BMPDTM** and the third **BMPDTM**

Next, after the first execution for Step (1b) and Step (1c) is performed, tube $T_1 = \{x_2^1 x_1^1, x_2^1 x_1^0\}$ and tube $T_2 = \{x_2^0 x_1^1, x_2^0 x_1^0\}$. This implies that the content of the *second* tape square for the tape in the first **BMDTM** in the second **BMPDTM** is written by its corresponding read-write head and is 1 ($x_2 = 1$), and the content of the *second* tape square for the tape in the second **BMDTM** in the second **BMPDTM** is written by its corresponding read-write head and is also 1 ($x_2 = 1$). Similarly, the content of the *second* tape square for the tape in the first **BMDTM** in the third **BMPDTM** is written by its corresponding read-write head and is 0 ($x_2 = 0$), and the content of the *second* tape square for the tape in the second **BMDTM** in the third **BMPDTM** is written by its corresponding read-write head and is also 0 ($x_2 = 0$). Figure 5.6 is used to show the current status of the execution environment to the second **BMPDTM** and the third **BMPDTM**.

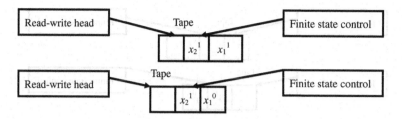

(a) The first **BMDTM** and the second **BMDTM** in the second **BMPDTM**.

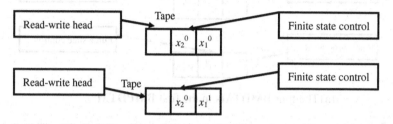

(b) The first **BMDTM** and the second **BMDTM** in the third **BMPDTM**.

Fig. 5.6 Schematic representation of the current status of the execution environment to the second **BMPDTM** and the third **BMPDTM**

Next, after the first execution for Step (1d) is finished, tube $T_0 = \{x_2^1 x_1^1, x_2^1 x_1^0, x_2^0 x_1^1, x_2^0 x_1^0\}$, tube $T_1 = \emptyset$ and tube $T_2 = \emptyset$. This is to say that the execution environment for those four bio-molecular deterministic one-tape Turing machines in the second **BMPDTM** and in the third **BMPDTM** becomes the first **BMPDTM**. Figure 5.7 is used to show the current status of the execution environment to the first **BMPDTM**. From Fig. 5.7, the contents to the four tapes in the execution environment of the first **BMPDTM** are not changed, the position of each read-write head and the status of each finite state control are also not changed.

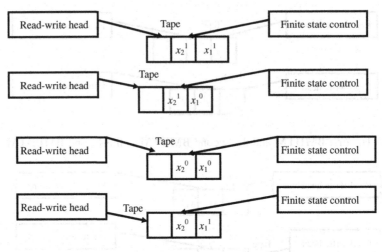

(a) The four **BMDTMs** in the first **BMPDTM**.

Fig. 5.7 Schematic representation of the current status of the execution environment to the first **BMPDTM**

Then, after the second execution of Step (1a) is implemented, tube $T_0 = \emptyset$, tube $T_1 = \{x_2^1 x_1^1, x_2^1 x_1^0, x_2^0 x_1^1, x_2^0 x_1^0\}$ and tube $T_2 = \{x_2^1 x_1^1, x_2^1 x_1^0, x_2^0 x_1^1, x_2^0 x_1^0\}$. This is to say that the four bio-molecular deterministic one-tape Turing machines in the execution environment of the first **BMPDTM** are all copied into the second **BMPDTM** and the third **BMPDTM**. Figures 5.8 and 5.9 are used to show the current status of the execution environment to the second **BMPDTM** and the third **BMPDTM**. From Fig. 5.8, the contents of four tapes in the execution environment of the second **BMPDTM** are, respectively, 11 ($x_2^1 x_1^1$), 10 ($x_2^1 x_1^0$), 01 ($x_2^0 x_1^1$) and 00 ($x_2^0 x_1^0$). Similarly, from Fig. 5.9, the contents of the four tapes in the execution environment of the third **BMPDTM** are also, respectively, 11 ($x_2^1 x_1^1$), 10 ($x_2^1 x_1^0$), 01 ($x_2^0 x_1^1$) and 00 ($x_2^0 x_1^0$). Simultaneously, from Figs. 5.8 and 5.9, the position and the status for each read-write head and each finite state control are reserved.

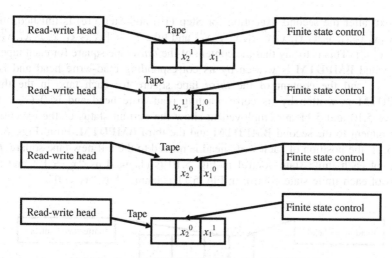

(a) The four **BMDTMs** in the second **BMPDTM.**

Fig. 5.8 Schematic representation of the current status of the execution environment to the second **BMPDTM**

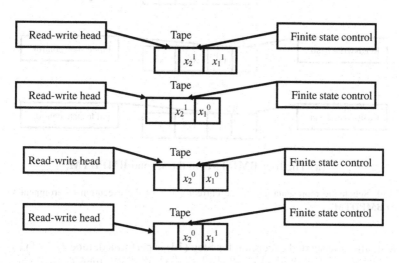

(a) The four **BMDTMs** in the third **BMPDTM.**

Fig. 5.9 Schematic representation of the current status of the execution environment to the third **BMPDTM**

Next, after the second execution for Step (1b) and Step (1c) is finished, tube $T_1 = \{x_3^1 x_2^1 x_1^1, x_3^1 x_2^1 x_1^0, x_3^1 x_2^0 x_1^1, x_3^1 x_2^0 x_1^0\}$ and tube $T_2 = \{x_3^0 x_2^1 x_1^1, x_3^0 x_2^1 x_1^0, x_3^0 x_2^0 x_1^1, x_3^0 x_2^0 x_1^0\}$. This is to say that the content of the *third* tape square for each tape in the second **BMPDTM** is written by its corresponding read-write head and is 1 ($x_3 = 1$), and the content of the *third* tape square for each tape in the third **BMPDTM** is written by its corresponding read-write head and is 0 ($x_3 = 0$). Figures 5.10 and 5.11 are employed to show the current status of the execution environment to the second **BMPDTM** and the third **BMPDTM**. From Figs. 5.10 and 5.11, the position of each write head is moved to the left new tape square, the status of each finite state control in Fig. 5.10 is changed as "$x_3 = 1$", and the status of each finite state control in Fig. 5.11 is changed as "$x_3 = 0$".

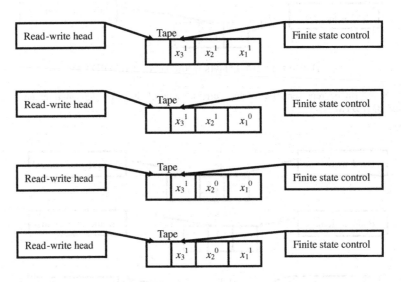

(a) The four **BMDTMs** in the second **BMPDTM**.

Fig. 5.10 Schematic representation of the current status of the execution environment to the second **BMPDTM**

Next, after the second execution for Step (1d) is performed, tube $T_0 = \{x_3^1 x_2^1 x_1^1, x_3^1 x_2^1 x_1^0, x_3^1 x_2^0 x_1^1, x_3^1 x_2^0 x_1^0, x_3^0 x_2^1 x_1^1, x_3^0 x_2^1 x_1^0, x_3^0 x_2^0 x_1^1, x_3^0 x_2^0 x_1^0\}$, tube $T_1 = \varnothing$ and tube $T_2 = \varnothing$. This implies that the execution environment for those bio-molecular deterministic one-tape Turing machines in the second **BMPDTM** and in the third **BMPDTM** becomes the first **BMPDTM**. Figure 5.12 is employed to show the current status of the execution environment to the first **BMPDTM**. From Fig. 5.12, the contents to the eight tapes in the execution environment of the first **BMPDTM** are not changed. Simultaneously, the position and the status for each read-write head and each finite state control are also not changed.

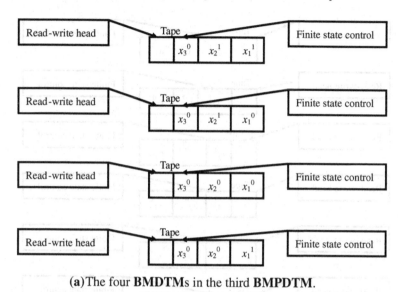

(a) The four **BMDTMs** in the third **BMPDTM**.

Fig. 5.11 Schematic representation of the current status of the execution environment to the third **BMPDTM**

Then, because Step (2) is the second loop and it's the lower and upper bounds for Step (2) are, respectively, two and three, Step (2a) will be executed two times. From the first execution, the second execution and the third execution of Step (2a), p_1^1, p_2^0, and p_3^0 are, respectively, appended into the head of each bit pattern in tube T_0. This implies that the contents of the fourth tape square, the fifth tape square and the sixth tape square for the eight tapes of the eight bio-molecular deterministic one-tape Turing machines in the first **BMPDTM** are written by the corresponding read-write head and are, subsequently, p_1^1, p_2^0, and p_3^0. Next, after the execution of Step (2b) is implemented, from each read-write head, z_0^0 is written into each tape. Figure 5.13 is used to show the current status of the execution environment to the first **BMPDTM**. From Fig. 5.13, the position of each read-write head is moved to the left new tape square, and the status of each finite state control is changed as "$z_0 = 0$".

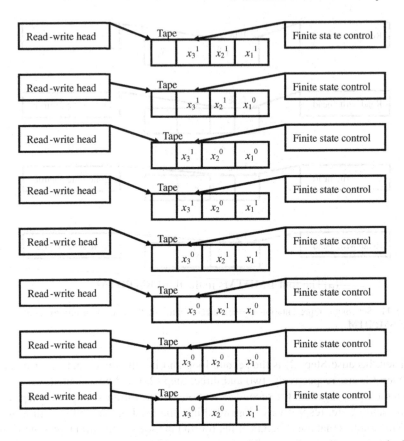

Fig. 5.12 Schematic representation of the current status of the execution environment to the first **BMPDTM**

Step (3) is the third loop and is used to perform an adder of n bits. When the first execution of Step (3a) is implemented, it invokes **Algorithm 5.1** that is used to perform an adder of one bit, **ParallelOneBitAdder**(T_0, k), in Sect. 5.2.1. The first parameter, tube T_0, is current the execution environment of the first **BMPDTM** and contains eight bio-molecular deterministic one-tape Turing machines (Fig. 5.13). It is regarded as an input tube of **Algorithm 5.1**. The value for the second parameter, k, is one and is also regarded as an input value of **Algorithm 5.1**.

When **Algorithm 5.1** is first called, seventeen tubes are all regarded as independent environments of seventeen bio-molecular parallel deterministic one-tape Turing machines. Tube T_0 is regarded as the first **BMPDTM** and tube T_k is regarded as the $(k + 1)$th **BMPDTM**. After the *first* execution of Step (1) in **Algorithm 5.1** is implemented, tube $T_0 = \varnothing$, tube $T_1 = \{z_0^0 p_3^0 p_2^0 p_1^1 x_3^1 x_2^1 x_1^1, z_0^0 p_3^0 p_2^0 p_1^1 x_3^1 x_2^0 x_1^1, z_0^0 p_3^0 p_2^0 p_1^1 x_3^1 x_2^1 x_1^0, z_0^0 p_3^0 p_2^0 p_1^1 x_3^1 x_2^0 x_1^0\}$ and tube $T_2 = \{z_0^0 p_3^0 p_2^0 p_1^1 x_3^1 x_2^1 x_1^0, z_0^0 p_3^0 p_2^0 p_1^1 x_3^1 x_2^0 x_1^0, z_0^0 p_3^0 p_2^0 p_1^1 x_3^0 x_2^1 x_1^0, z_0^0 p_3^0 p_2^0 p_1^1 x_3^0 x_2^0 x_1^0\}$. This implies that

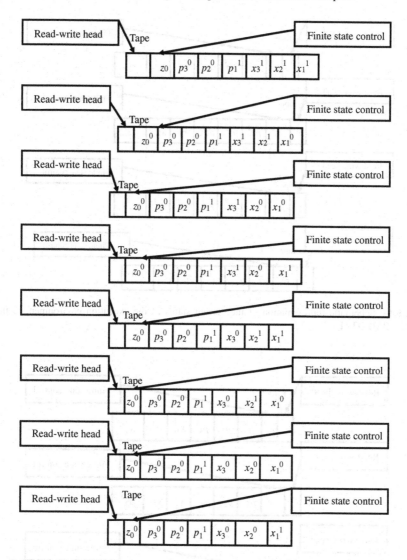

Fig. 5.13 Schematic representation of the current status of the execution environment to the first **BMPDTM**

the new execution environments for four bio-molecular deterministic one-tape Turing machines with the content of tape square, "x_1^1", and other bio-molecular deterministic one-tape Turing machines with the content of tape square, "x_1^0" are, respectively, the second **BMPDTM** and the third **BMPDTM**. Figures 5.14 and 5.15 are applied to show the results. From Figs. 5.14 and 5.15, the position of each read-write head is reserved, and the status of each finite state control is reserved.

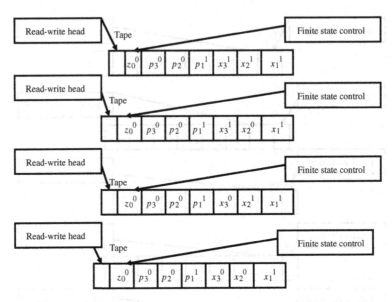

Fig. 5.14 Schematic representation of the current status of the execution environment to the second **BMPDTM**

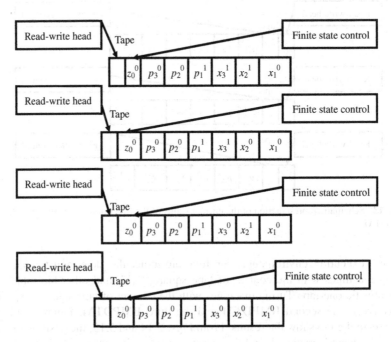

Fig. 5.15 Schematic representation of the current status of the execution environment to the third **BMPDTM**

Next, after the *first* execution of Step (2) in **Algorithm 5.1** is implemented, tube $T_1 = \varnothing$, tube $T_3 = \{z_0^0\, p_3^0\, p_2^0\, p_1^1\, x_3^1\, x_2^1\, x_1^1,\ z_0^0\, p_3^0\, p_2^0\, p_1^1\, x_3^1\, x_2^0\, x_1^1,\ z_0^0\, p_3^0\, p_2^0\, p_1^1\, x_3^0\, x_2^1\, x_1^1,\ z_0^0\, p_3^0\, p_2^0\, p_1^1\, x_3^0\, x_2^0\, x_1^1\}$ and tube $T_4 = \varnothing$. This is to say that the new execution environment for four bio-molecular deterministic one-tape Turing machines with the content of tape square, "p_1^1" is the fourth **BMPDTM**. Figure 5.16 is applied to show the execution environment for the fourth **BMPDTM**. From Fig. 5.16, the position of each read-write head is reserved, and the status of each finite state control is reserved.

Next, after the *first* execution of Step (3) in **Algorithm 5.1** is implemented, tube $T_2 = \varnothing$, tube $T_5 = \{z_0^0\, p_3^0\, p_2^0\, p_1^1\, x_3^1\, x_2^1\, x_1^0,\ z_0^0\, p_3^0\, p_2^0\, p_1^1\, x_3^1\, x_2^0\, x_1^0,\ z_0^0\, p_3^0\, p_2^0\, p_1^1\, x_3^0\, x_2^1\, x_1^0,\ z_0^0\, p_3^0\, p_2^0\, p_1^1\, x_3^0\, x_2^0\, x_1^0\}$ and tube $T_6 = \varnothing$. Figure 5.17 is applied to show the execution environment for the sixth **BMPDTM**. From Fig. 5.17, the position of each read-write head is reserved, and the status of each finite state control is reserved.

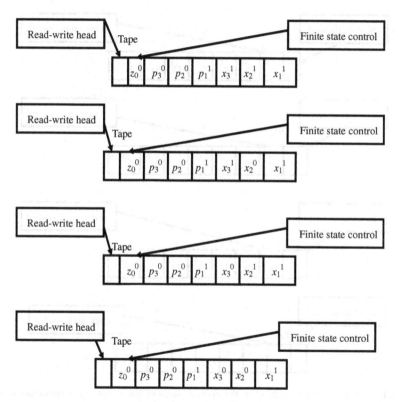

Fig. 5.16 Schematic representation of the current status of the execution environment to the fourth **BMPDTM**

After the first execution for Steps (4) through (7) is performed, tube $T_8 = \{z_0^0 p_3^0$
$p_2^0 p_1^1 x_3^1 x_2^1 x_1^1, z_0^0 p_3^0 p_2^0 p_1^1 x_3^1 x_2^1 x_1^0, z_0^0 p_3^0 p_2^0 p_1^1 x_3^1 x_2^0 x_1^1, z_0^0 p_3^0 p_2^0 p_1^1 x_3^0 x_2^0 x_1^1\}$, tube
$T_{12} = \{z_0^0 p_3^0 p_2^0 p_1^1 x_3^1 x_2^1 x_1^1, z_0^0 p_3^0 p_2^0 p_1^1 x_3^1 x_2^1 x_1^1, z_0^0 p_3^0 p_2^0 p_1^1 x_3^1 x_2^1 x_1^1, z_0^0 p_3^0 p_2^0 p_1^1 x_3^1 x_2^0$
$x_1^0\}$, tube $T_7 = \varnothing$, tube $T_9 = \varnothing$, tube $T_{10} = \varnothing$, tube $T_{11} = \varnothing$, tube $T_{13} = \varnothing$, and
tube $T_{14} = \varnothing$. Figures 5.18 and 5.19 are, respectively, used to show the results.
From Figs. 5.18 and 5.19, the position of each read-write head is reserved, and the
status of each finite state control is reserved.

Next, after the first execution for Steps (8a) (9a) (10a) (11a) (12a) (13a) (14a)
and (15a) is performed, the returned result from Steps (9a) and (13a) is "yes", and
other returned result is "no". Therefore, after the first execution for Steps (9) and
(13) is finished, tube $T_8 = \{z_1^1 y_1^0 z_0^0 p_3^0 p_2^0 p_1^1 x_3^1 x_2^1 x_1^1, z_1^1 y_1^0 z_0^0 p_3^0 p_2^0 p_1^1 x_3^1 x_2^1 x_1^1, z_1^1 y_1^0$
$z_0^0 p_3^0 p_2^0 p_1^1 x_3^1 x_2^1 x_1^1, z_1^1 y_1^0 z_0^0 p_3^0 p_2^0 p_1^1 x_3^1 x_2^1 x_1^1\}$, tube $T_{12} = \{z_1^0 y_1^1 z_0^0 p_3^0 p_2^0 p_1^1 x_3^1 x_2^1 x_1^1,$
$z_1^0 y_1^1 z_0^0 p_3^0 p_2^0 p_1^1 x_3^1 x_2^1 x_1^0, z_1^0 y_1^1 z_0^0 p_3^0 p_2^0 p_1^1 x_3^1 x_2^1 x_1^0, z_1^0 y_1^1 z_0^0 p_3^0 p_2^0 p_1^1 x_3^0 x_2^0 x_1^0\}$. Finally,

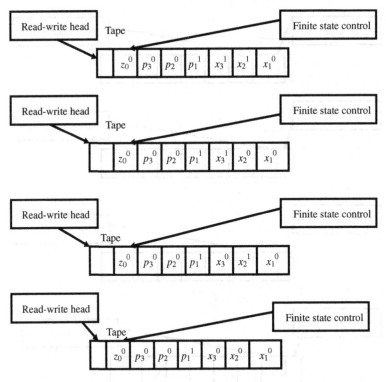

Fig. 5.17 Schematic representation of the current status of the execution environment to the sixth **BMPDTM**

after the first execution for Step (16) is performed, tube $T_0 = \{z_1^1 \, y_1^0 \, z_0^0 \, p_3^0 \, p_2^0 \, p_1^1 \, x_3^1$
$x_2^1 \, x_1^1, \, z_1^1 \, y_1^0 \, z_0^0 \, p_3^0 \, p_2^0 \, p_1^0 \, x_3^1 \, x_2^1 \, x_1^1, \, z_1^1 \, y_1^0 \, z_0^0 \, p_3^0 \, p_2^0 \, p_1^1 \, x_3^0 \, x_2^1 \, x_1^1, \, z_1^0$
$y_1^1 \, z_0^0 \, p_3^0 \, p_2^0 \, p_1^1 \, x_3^1 \, x_2^1 \, x_1^1, \, z_1^1 \, y_1^1 \, z_0^0 \, p_3^0 \, p_2^0 \, p_1^1 \, x_3^1 \, x_2^1 \, x_1^1, \, z_1^1 \, y_1^1 \, z_0^0 \, p_3^0$
$p_2^0 \, p_1^1 \, x_3^0 \, x_2^0 \, x_1^0\}$ and other tubes become all empty tubes. Figures 5.20, 5.21 and 5.22
are used to show the result. From Figs. 5.20 and 5.21, the position of each read-
write head is moved to the left new tape square, from Fig. 5.20, the status of each
finite state control is changed as "$z_1 = 1$", and from Fig. 5.21, the status of each
finite state control is changed as "$z_1 = 0$".

After the execution of the *first* time for each operation in **Algorithm 5.1** is
performed, the addition for the first bit for each augend and its corresponding
addend is also finished. Then, when the *second* execution of Step (3a) is implemented, it again invokes **Algorithm 5.1**. The first parameter, tube T_0, is current the

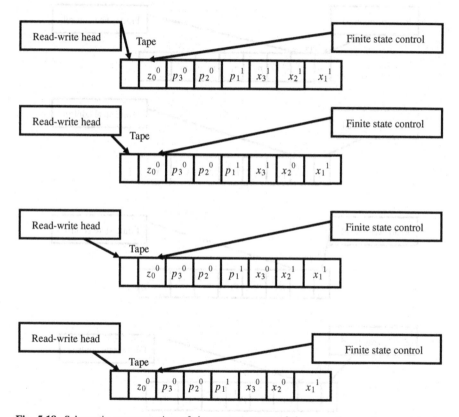

Fig. 5.18 Schematic representation of the current status of the execution environment to the
ninth **BMPDTM**

execution environment of the first **BMPDTM** and contains eight bio-molecular deterministic one-tape Turing machines (Fig. 5.22). It is regarded as an input tube of **Algorithm 5.1**. The value for the second parameter, k, is *two* and is also regarded as an input value of **Algorithm 5.1**. After the execution of the *second* time for each operation in **Algorithm 5.1** is performed, tube $T_0 = \{z_2^1 y_2^0 z_1^1 y_1^0 z_0^0 p_3^0$ $p_2^0 p_1^1 x_3^1 x_2^1 x_1^1, z_2^0 y_2^1 z_1^0 y_1^1 z_0^0 p_3^0 p_2^0 p_1^1 x_3^1 x_2^1 x_1^0, z_2^0 y_2^1 z_1^0 y_1^1 z_0^0 p_3^0 p_2^0 p_1^1 x_3^1 x_2^0 x_1^1, z_2^0 y_2^0 z_1^1 y_1^1$ $z_0^0 p_3^0 p_2^0 p_1^1 x_3^1 x_2^0 x_1^0, z_2^0 y_2^1 z_1^0 y_1^0 z_0^1 p_3^0 p_2^0 p_1^0 x_3^1 x_2^1 x_1^1, z_2^0 y_2^0 z_1^1 y_1^0 z_0^1 p_3^0 p_2^0 p_1^0 x_3^1 x_2^1 x_1^0, z_2^0$ $y_2^0 z_1^1 y_1^0 z_0^0 p_3^0 p_2^0 p_1^1 x_3^0 x_2^0 x_1^1, z_2^0 y_2^0 z_1^0 y_1^1 z_0^0 p_3^0 p_2^0 p_1^1 x_3^0 x_2^0 x_1^0\}$ and other tubes become all empty tubes. Figure 5.23 is used to show the result.

After the execution of the *second* time for each operation in **Algorithm 5.1** is performed, the addition for the *second* bit for each augend and its corresponding

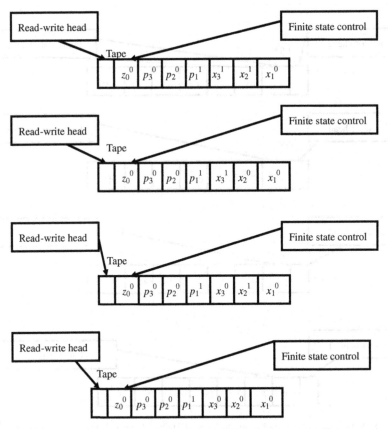

Fig. 5.19 Schematic representation of the current status of the execution environment to the 13th **BMPDTM**

addend is also finished. Next, when the *third* execution of Step (3a) is implemented, it again invokes **Algorithm 5.1**. The first parameter, tube T_0, is current the execution environment of the first **BMPDTM** and contains eight bio-molecular deterministic one-tape Turing machines (Fig. 5.23). It is regarded as an input tube of **Algorithm 5.1**. The value for the second parameter, k, is *three* and is also regarded as an input value of **Algorithm 5.1**. After the execution of the *third* time for each operation in **Algorithm 5.1** is performed, tube $T_0 = \{z_3^1\, y_3^0\, z_2^1\, y_2^0\, z_1^1\, y_1^0\, z_0^0\, p_3^0\, p_2^0\, p_1^1\, x_3^1\, x_2^1\, x_1^1,\, z_3^0\, y_3^1\, z_2^0\, y_2^1\, z_1^0\, y_1^1\, z_0^0\, p_3^0\, p_2^0\, p_1^1\, x_3^1\, x_2^1\, x_1^0,\, z_3^0\, y_3^1\, z_2^0\, y_2^1\, z_1^1\, y_1^0\, z_0^0\, p_3^0\, p_2^0\, p_1^1\, x_3^1\, x_2^0\, x_1^1,$
$z_3^0\, y_3^1\, z_2^0\, y_2 z_1^0\, y_1^1\, z_0^0\, p_3^0\, p_2^0\, p_1^1\, x_3^1\, x_2^0\, x_1^0,\, z_3^0\, y_3^0\, z_2^0\, y_2^1\, z_1^1\, y_1^0\, z_0^0\, p_3^0\, p_2^0\, p_1^1\, x_3^0\, x_2^1\, x_1^1,\, z_3^0\, y_3^1\, z_2^0\, y_2 z_1^1\, y_1^0$
$z_0^0\, p_3^0\, p_2^0\, p_1^1\, x_3^0\, x_2^1\, x_1^0,\, z_3^0\, y_3^0\, z_2^0\, y_2 z_1^1\, y_1^0\, z_0^0\, p_3^0\, p_2^0\, p_1^1\, x_3^0\, x_2^0\, x_1^1,\, z_3^0\, y_3^0\, z_2^0\, y_2^1\, z_1^0\, y_1^1\, z_0^0\, p_3^0\, p_2^0\, p_1^1\, x_3^0$
$x_2^0\, x_1^0\}$ and other tubes become all empty tubes. Figure 5.24 is used to show the result.

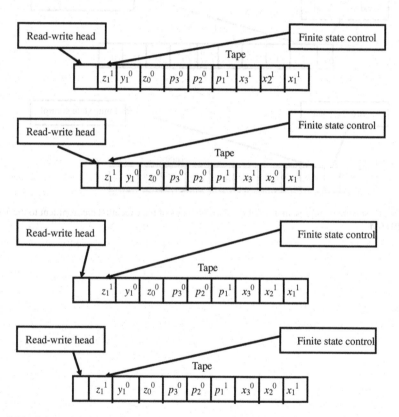

Fig. 5.20 Schematic representation of the current status of the execution environment to the ninth **BMPDTM**

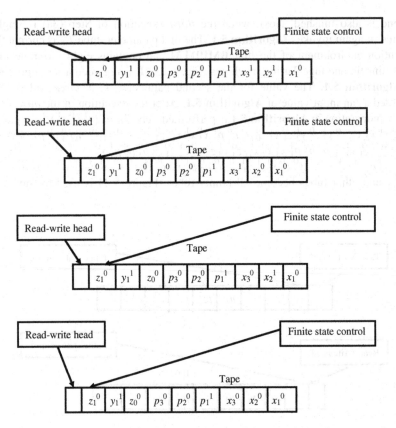

Fig. 5.21 Schematic representation of the current status of the execution environment to the 13th BMPDTM

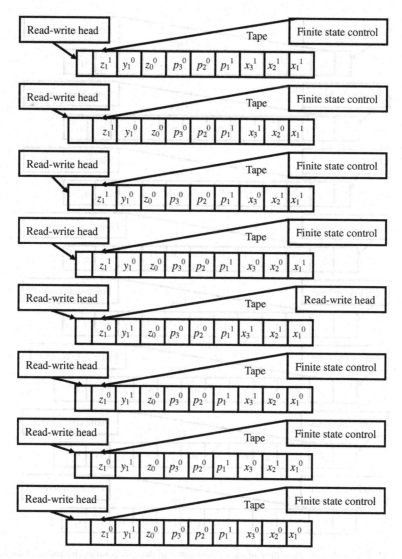

Fig. 5.22 Schematic representation of the current status of the execution environment to the first **BMPDTM**

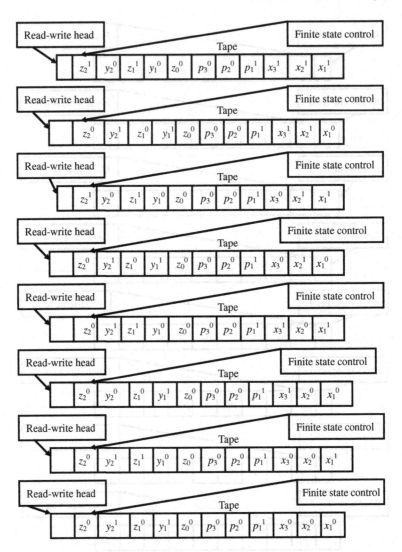

Fig. 5.23 Schematic representation of the current status of the execution environment to the first **BMPDTM**

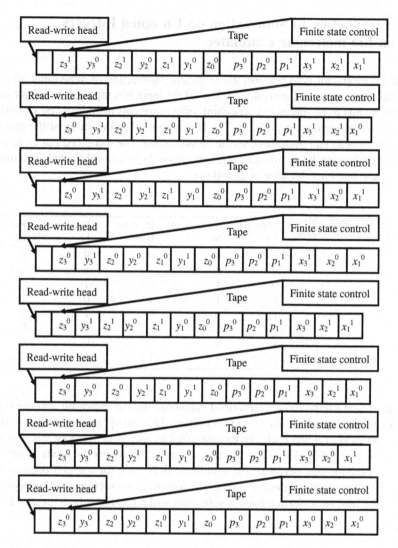

Fig. 5.24 Schematic representation of the current status of the execution environment to the first **BMPDTM**

5.3 Introduction to Subtraction on Unsigned Integers
   on Bio-molecular Computer

A one-bit subtractor is to perform the arithmetic subtraction of three input bits. It contains three inputs and two outputs. Two of the input bits represent minuend and subtrahend bits to be subtracted. The third input represents the borrow bit from the previous lower significant position. The first output gives the value of the difference for minuend and subtrahend bits to be subtracted. The second output gives the value of the borrow bit to minuend and subtrahend bits to be subtracted. The truth table of the one-bit subtractor is as follows:

Table 5.2 The truth table of a one-bit subtractor is shown

Minuend bit	Subtrahend bit	Previous borrow bit	Difference bit	Borrow bit
0	0	0	0	0
0	0	1	1	1
0	1	0	1	1
0	1	1	0	1
1	0	0	1	0
1	0	1	0	0
1	1	0	0	0
1	1	1	1	1

The one-bit subtractor just described calculates the difference bit and the borrow bit for two input bits and a previous borrow. The first operand of n-bits and the second operand of n-bits each can be subtracted by means of finishing this one-bit subtractor of n times. This is to say that a parallel subtractor of n-bits is also a Boolean function that performs the arithmetic difference for the first operand of n-bits and the second operand of n-bits. The following subsections will be applied to describe how to perform a one-bit subtractor and a parallel subtractor of n-bits.

5.3.1 The Construction of a Parallel One-bit Subtractor
   on Unsigned Integers on Bio-molecular Computer

Suppose that two one-bit binary numbers, m_k and d_k, are employed to represent the first input of a one-bit subtractor, and the first output of a one-bit subtractor for $1 \leq k \leq n$, respectively, a one-bit binary number, s_k, is used to represent the second input of a one-bit subtractor, and two one-bit binary numbers, b_k and b_{k-1}, are applied to represent the second output and the third input of a one-bit subtractor, respectively. For the sake of convenience, assume that m_k^1 contains the value of m_k to be 1 and m_k^0 contains the value of m_k to be 0. Similarly, also suppose that d_k^1 contains the value of d_k to be 1 and d_k^0 contains the value of d_k to be 0. Assume that s_k^1 denotes the value of s_k to be 1 and $s_k{}^0$ defines the value of s_k to be 0. Similarly, also suppose that b_k^1 denotes the value of b_k to be 1 and $b_k{}^0$ defines the value of b_k to be 0. Assume that b_{k-1}^1 contains the value of b_{k-1} to be 1 and b_{k-1}^0 contains the value of b_{k-1} to be 0. The following algorithm is proposed to perform the Boolean function of a parallel one-bit subtractor.

Algorithm 5.3: ParallelOneBitSubtractor(T_0, k)

(1) $T_1 = +(T_0, m_k^1)$ and $T_2 = -(T_0, m_k^1)$.

(2) $T_3 = +(T_1, s_k^1)$ and $T_4 = -(T_1, s_k^1)$.

(3) $T_5 = +(T_2, s_k^1)$ and $T_6 = -(T_2, s_k^1)$.

(4) $T_7 = +(T_3, b_{k-1}^1)$ and $T_8 = -(T_3, b_{k-1}^1)$.

(5) $T_9 = +(T_4, b_{k-1}^1)$ and $T_{10} = -(T_4, b_{k-1}^1)$.

(6) $T_{11} = +(T_5, b_{k-1}^1)$ and $T_{12} = -(T_5, b_{k-1}^1)$.

(7) $T_{13} = +(T_6, b_{k-1}^1)$ and $T_{14} = -(T_6, b_{k-1}^1)$.

(8a) If (Detect(T_7) $= =$ "yes") then

    (8a1) Append-head(T_7, d_k^1) and Append-head(T_7, b_k^1).

  EndIf

(9a) If (Detect(T_8) $= =$ "yes") then

    (9a1) Append-head(T_8, d_k^0) and Append-head(T_8, b_k^0).

  EndIf

(10a) If (Detect(T_9) $= =$ "yes") then

    (10a1) Append-head(T_9, d_k^0) and Append-head(T_9, b_k^0).

  EndIf

(11a) If (Detect(T_{10}) $= =$ "yes") then

    (11a1) Append-head(T_{10}, d_k^1) and Append-head(T_{10}, b_k^0).

  EndIf

(12a) If (Detect(T_{11}) $= =$ "yes") then

    (12a1) Append-head(T_{11}, d_k^0) and Append-head(T_{11}, b_k^1).

  EndIf

(13a) If (Detect(T_{12}) $= =$ "yes") then

    (13a1) Append-head(T_{12}, d_k^1) and Append-head(T_{12}, b_k^1).

  EndIf

(14a) If (Detect(T_{13}) $= =$ "yes") then

    (14a1) Append-head(T_{13}, d_k^1) and Append-head(T_{13}, b_k^1).

  EndIf

(15a) If (Detect(T_{14}) $= =$ "yes") then

    (15a1) Append-head(T_{14}, d_k^0) and Append-head(T_{14}, b_k^0).

  EndIf

(16) $T_0 = \cup(T_7, T_8, T_9, T_{10}, T_{11}, T_{12}, T_{13}, T_{14})$.

EndProcedure

Lemma 5-3: *The algorithm,* **ParallelOneBitSubtractor**(T_0, k), *can be applied to finish the function of a parallel one-bit subtractor.*

Proof The algorithm, **ParallelOneBitSubtractor**(T_0, k), is implemented by means of the *extract, append-head* and *merge* operations. Steps (1) through (7) employ the *extract* operations to form some different test tubes including different inputs (T_1 to T_{14}). That is, T_1 includes all of the inputs that have $m_k = 1$, T_2 includes all of the inputs that have $m_k = 0$, T_3 includes those that have $m_k = 1$ and $s_k = 1$, T_4 includes those that have $m_k = 1$ and $s_k = 0$, T_5 includes those that have $m_k = 0$ and $s_k = 1$, T_6 includes those that have $m_k = 0$ and $s_k = 0$, T_7 includes those that have $m_k = 1$, $s_k = 1$ and $b_{k-1} = 1$, T_8 includes those that have $m_k = 1$, $s_k = 1$ and $b_{k-1} = 0$, T_9 includes those that have $m_k = 1$, $s_k = 0$ and $b_{k-1} = 1$, T_{10} consists of those that have $m_k = 1$, $s_k = 0$ and $b_{k-1} = 0$, T_{11} includes those that have $m_k = 0$, $s_k = 1$ and $b_{k-1} = 1$, T_{12} includes those that have $m_k = 0$, $s_k = 1$ and $b_{k-1} = 0$, T_{13} includes those that have $m_k = 0$, $s_k = 0$ and $b_{k-1} = 1$, and finally, T_{14} consists of those that have $m_k = 0$, $s_k = 0$ and $b_{k-1} = 0$. Having finished Steps (1) through (7), this implies that eight different inputs of a one-bit subtractor as shown in Table 5.2 were poured into tubes T_7 through T_{14}, respectively.

Next, Steps (8a) (9a) (10a) (11a) (12a) (13a) (14a) and (15a) are used to detect whether there are inputs from tubes T_7 through T_{14}. If a "yes" is returned from each operation, then the corresponding steps (8a1) through (15a1) use the *append-head* operations to append d_k^1 or d_k^0, and b_k^1 or b_k^0 onto the head of every input in the corresponding test tubes. After finishing Steps (8a1) through (15a1), we can say that eight different outputs of a one-bit subtractor in Table 5.2 are appended into tubes T_7 through T_{14}. Finally, the execution of Step (16) applies the *merge* operation to pour tubes T_7 through T_{14} into tube T_0. Tube T_0 contains the result finishing the subtraction of a bit.   ∎

5.3.2 The Construction of a Parallel N-*bits Subtractor on Unsigned Integers on Bio-molecular Computer*

The parallel one-bit subtractor described in Sect. 5.3.1 is applied to figure out the difference and borrow for two input bits and a previous borrow. It directly employs the truth table (Table 5.2 in Sect. 5.3.1) to finish subtraction of a bit. The main distinction between one-bit subtractor on bio-molecular computer and a full subtractor of one-bit on Turing's machines is that the Karnaugh map and basic logic gates on Turing's machines are not used to implement subtraction of a bit on bio-molecular computer. Similarly, a parallel subtractor of n-bits is also directly to perform the arithmetic difference for the first operand of n-bits and the second operand of n-bits by means of finishing this one-bit subtractor of n times. The following algorithm is offered to perform the arithmetic difference for a parallel subtractor of n-bits.

Algorithm 5.4: BinaryParallelSubtractor(T_0)

(1) Append-head(T_1, m_1^1).

(2) Append-head(T_2, m_1^0).

(3) $T_0 = \cup(T_1, T_2)$.

(4) **For** $k = 2$ **to** n

    (4a) Amplify(T_0, T_1, T_2).

    (4b) Append-head(T_1, m_k^1).

    (4c) Append-head(T_2, m_k^0).

    (4d) $T_0 = \cup(T_1, T_2)$.

EndFor

(5) **For** $k = 1$ **to** n

    (5a) Append-head(T_0, s_k).

EndFor

(6) Append-head(T_0, b_0^0).

(7) **For** $k = 1$ **to** n

    (7a) **ParallelOneBitSubtractor(T_0, k).**

EndFor

EndAlgorithm

Lemma 5-4: *The algorithm,* **BinaryParallelSubtractor**(T_0), *can be used to finish the Boolean function to a binary parallel subtractor of n bits.*

Proof Steps (1) through (4d) are mainly used to construct solution space of 2^n unsigned integers (the range of values for them is from 0 to $2^n - 1$). After they are finished, tube T_0 includes those inputs encoding 2^n unsigned integers. Step (5) is the second loop and is applied to generate subtrahend bits for a subtractor of n bits. On each execution of Step (5a), it applies the "Append-head" operation to append the value "0" or "1" for the kth bit of a subtrahend into the head of each bit pattern in tube T_0. When the operation for a subtractor of n bits, the least significant position for the minuend and the subtrahend is subtracted, the previous borrow bit must be 0. On the execution of Step (6), it uses the "Append-head" operation to append the value 0 of the previous borrow bit, b_0, onto the end of every bit pattern in T_0.

Next, Step (7) is the main loop and is mainly used to perform the Boolean function of a binary parallel subtractor of n bits. Each execution of Step (7a) calls **ParallelOneBitSubtractor**(T_0, k) in Sect. 5.3.1 to compute the arithmetic difference of one bit for the minuend and the subtrahend. Repeat execution of Step (7a) until the most significant bit for the minuend and the subtrahend is processed. Tube T_0 contains the result performing the arithmetic difference for the first operand of n-bits and the second operand of n-bits. ∎

5.3.3 The Power for a Parallel N-bits Subtractor
on Unsigned Integers on Bio-molecular Computer

Consider that eight values for an unsigned integer of three bits are, subsequently, $000(0_{10})$ $(m_3^0\ m_2^0\ m_1^0)$, $001(1_{10})$ $(m_3^0\ m_2^0\ m_1^1)$, $010(2_{10})$ $(m_3^0\ m_2^1\ m_1^0)$, $011(3_{10})$ $(m_3^0\ m_2^1\ m_1^1)$, $100(4_{10})$ $(m_3^1\ m_2^0\ m_1^0)$, $101(5_{10})$ $(m_3^1\ m_2^0\ m_1^1)$, $110(6_{10})$ $(m_3^1\ m_2^1\ m_1^0)$ and $111(7_{10})$ $(m_3^1\ m_2^1\ m_1^1)$. We want to simultaneously subtract $001(1_{10})$ $(s_3^0\ s_2^0\ s_1^1)$ from those eight values. **Algorithm 5.4, BinaryParallelSubtractor(T_0)**, can be used to implement the task. Tube T_0 is an empty tube and is regarded as an input tube of **Algorithm 5.4**. From **Definition 5.2**, the input tube T_0 is regarded as the execution environment of the first **BMPDTM**. Similarly, tubes T_1 and T_2 used in **Algorithm 5.4** also are regarded, respectively, as the execution environment of the second **BMPDTM** and the execution environment of the third **BMPDTM**.

Steps (1) through (4d) in **Algorithm 5.4** are used to construct a **BMPDTM** with eight bio-molecular deterministic one-tape Turing machines. After the execution for Step (1) and Step (2) of **Algorithm 5.4** is finished, tube $T_1 = \{m_1^1\}$ and tube $T_2 = \{m_1^0\}$. This is to say that a **BMDTM** in the second **BMPDTM** and in the third **BMPDTM** is constructed. Figure 5.25 is applied to explain the current status of the execution environment to the second **BMPDTM** and the third **BMPDTM**. From Fig. 5.25, the content of the first tape square for the tape in the first **BMDTM** in the second **BMPDTM** is written by its corresponding read-write head and is 1 ($m_1 = 1$), and the content of the first tape square for the tape in the first **BMDTM** in the third **BMPDTM** is written by its corresponding read-write head and is 0 ($m_1 = 0$).

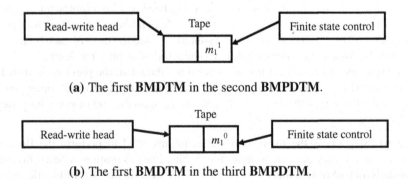

(a) The first **BMDTM** in the second **BMPDTM**.

(b) The first **BMDTM** in the third **BMPDTM**.

Fig. 5.25 Schematic representation of the current status of the execution environment to the second **BMPDTM** and the third **BMPDTM**

Next, after the execution for Step (3) of **Algorithm 5.4** is implemented, tube $T_0 = \{m_1^1,\ m_1^0\}$, tube $T_1 = \varnothing$ and tube $T_2 = \varnothing$. This implies that the execution environment for the first **BMDTM** in the second **BMPDTM** and the first

BMDTM in the third **BMPDTM** becomes the first **BMPDTM**. Figure 5.26 is used to show the current status of the execution environment to the first **BMPDTM**. From Fig. 5.26, the contents to the two tapes in the execution environment of the first **BMPDTM** are not changed, and the position of each read-write head and the status of each finite state control are also not changed.

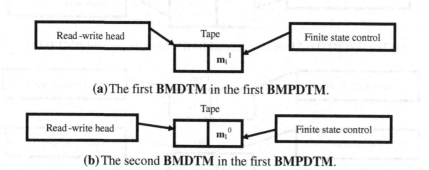

(a) The first **BMDTM** in the first **BMPDTM**.

(b) The second **BMDTM** in the first **BMPDTM**.

Fig. 5.26 Schematic representation of the current status of the execution environment to the first **BMPDTM**

Step (4) is the first loop in **Algorithm 5.4**, the upper bound (n) is three because the number of bits for representing those eight values is three. Therefore, after the first execution of Step (4a) is performed, tube $T_0 = \varnothing$, tube $T_1 = \{m_1^1, m_1^0\}$ and tube $T_2 = \{m_1^1, m_1^0\}$. This is to say that the first **BMDTM** and the second **BMDTM** in the execution environment of the first **BMPDTM** are both copied into the second **BMPDTM** and the third **BMPDTM**. Figure 5.27 is employed to explain the current status of the execution environment to the second **BMPDTM** and the third **BMPDTM**. From Fig. 5.27, the contents of the first tape square for the corresponding tape of the first **BMDTM** and the corresponding tape of the second **BMDTM** in the execution environment of the second **BMPDTM** are, respectively, 1 ($m_1 = 1$) and 0 ($m_1 = 0$). The contents of the first tape square for the corresponding tape of the first **BMDTM** and the corresponding tape of the second **BMDTM** in the execution environment of the third **BMPDTM** are also, respectively, 0 ($m_1 = 0$) and 1 ($m_1 = 1$). From Fig. 5.27, it is pointed out that four bio-molecular deterministic one-tape Turing machines are generated.

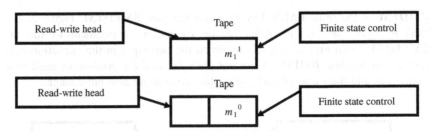

(a) The first **BMDTM** and the second BMDTM in the second **BMPDTM**.

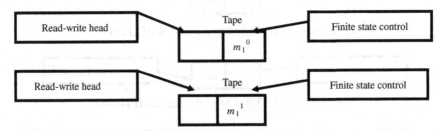

(b) The first **BMDTM** and the second **BMDTM** in the third **BMPDTM**.

Fig. 5.27 Schematic representation of the current status of the execution environment to the second **BMPDTM** and the third **BMPDTM**

Next, after the first execution for Step (4b) and Step (4c) of **Algorithm 5.4** is implemented, tube $T_1 = \{m_2^1 \, m_1^1, \, m_2^1 \, m_1^0\}$ and tube $T_2 = \{m_2^0 \, m_1^1, \, m_2^0 \, m_1^0\}$. This implies that the content of the *second* tape square for the tape in the first **BMDTM** in the second **BMPDTM** is written by its corresponding read-write head and is 1 ($m_2 = 1$), and the content of the *second* tape square for the tape in the second **BMDTM** in the second **BMPDTM** is written by its corresponding read-write head and is also 1 ($m_2 = 1$). Similarly, the content of the *second* tape square for the tape in the first **BMDTM** in the third **BMPDTM** is written by its corresponding read-write head and is 0 ($m_2 = 0$), and the content of the *second* tape square for the tape in the second **BMDTM** in the third **BMPDTM** is written by its corresponding read-write head and is also 0 ($m_2 = 0$). Figure 5.28 is used to show the current status of the execution environment to the second **BMPDTM** and the third **BMPDTM**. From Fig. 5.28, the position of each read-write head is moved to the left new tape square, and the status of each finite state control is changed as "$m_2 = 1$" and "$m_2 = 0$".

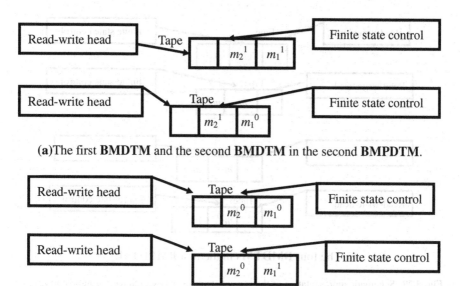

(a) The first **BMDTM** and the second **BMDTM** in the second **BMPDTM**.

(b) The first BMDTM and the second **BMDTM** in the third **BMPDTM**.

Fig. 5.28 Schematic representation of the current status of the execution environment to the second **BMPDTM** and the third **BMPDTM**

Next, after the first execution for Step (4d) of **Algorithm 5.4** is implemented, tube $T_0 = \{m_2^1 m_1^1, m_2^1 m_1^0, m_2^0 m_1^1, m_2^0 m_1^0\}$, tube $T_1 = \varnothing$ and tube $T_2 = \varnothing$. This is to say that the execution environment for those bio-molecular deterministic one-tape Turing machines in the second **BMPDTM** and in the third **BMPDTM** becomes the first **BMPDTM**. Figure 5.29 is applied to explain the current status of the execution environment to the first **BMPDTM**. From Fig. 5.29, the contents to the four tapes in the execution environment of the first **BMPDTM** are not changed, and the position of each read-write head and the status of each finite state control are also not changed.

Then, after the second execution of Step (4a) is implemented, tube $T_0 = \varnothing$, tube $T_1 = \{m_2^1 m_1^1, m_2^1 m_1^0, m_2^0 m_1^1, m_2^0 m_1^0\}$ and tube $T_2 = \{m_2^1 m_1^1, m_2^1 m_1^0, m_2^0 m_1^1, m_2^0 m_1^0\}$. This is to say that the four bio-molecular deterministic one-tape Turing machines in the execution environment of the first **BMPDTM** are all copied into the second **BMPDTM** and the third **BMPDTM**. Figures 5.30 and 5.31 are used to show the current status of the execution environment to the second **BMPDTM** and the third **BMPDTM**. From Fig. 5.30, the contents of four tapes in the execution environment of the second **BMPDTM** are, respectively, 11 ($m_2^1 m_1^1$), 10 ($m_2^1 m_1^0$), 01 ($m_2^0 m_1^1$) and 00 ($m_2^0 m_1^0$). Similarly, from Fig. 5.31, the contents of the four tapes in the execution environment of the third **BMPDTM** are also, respectively, 11 ($m_2^1 m_1^1$), 10 ($m_2^1 m_1^0$), 01 ($m_2^0 m_1^1$) and 00 ($m_2^0 m_1^0$).

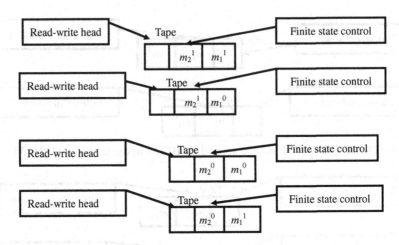

(a) The four **BMDTMs** in the first **BMPDTM**.

Fig. 5.29 Schematic representation of the current status of the execution environment to the first **BMPDTM**

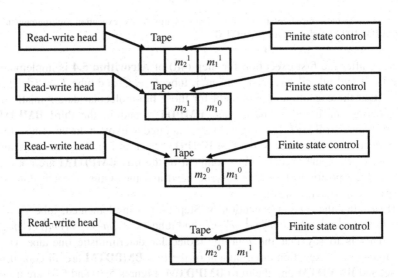

(a) The four **BMDTMs** in the second **BMPDTM**.

Fig. 5.30 Schematic representation of the current status of the execution environment to the second **BMPDTM**

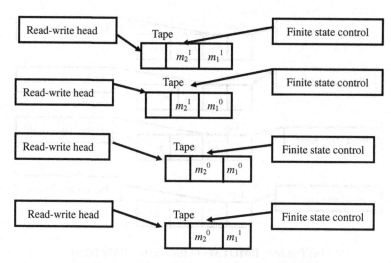

(a) The four **BMDTMs** in the third **BMPDTM**.

Fig. 5.31 Schematic representation of the current status of the execution environment to the third **BMPDTM**

Next, after the second execution for Step (4b) and Step (4c) of **Algorithm 5.4** is performed, tube $T_1 = \{m_3^1\, m_2^1\, m_1^1,\ m_3^1\, m_2^1\, m_1^0,\ m_3^1\, m_2^0\, m_1^1,\ m_3^1\, m_2^0\, m_1^0\}$ and tube $T_2 = \{m_3^0\, m_2^1\, m_1^1,\ m_3^0\, m_2^1\, m_1^0,\ m_3^0\, m_2^0\, m_1^1,\ m_3^0\, m_2^0\, m_1^0\}$. This is to say that the content of the *third* tape square for each tape in the second **BMPDTM** is written by its corresponding read-write head and is 1 ($m_3 = 1$), and the content of the *third* tape square for each tape in the third **BMPDTM** is written by its corresponding read-write head and is 0 ($m_3 = 0$). Figures 5.32 and 5.33 are employed to show the current status of the execution environment to the second **BMPDTM** and the third **BMPDTM**. From Figs. 5.32 and 5.33, the position of each read-write head is moved to the left new tape square, and the status of each finite state control is changed as "$m_3 = 1$" and "$m_3 = 0$".

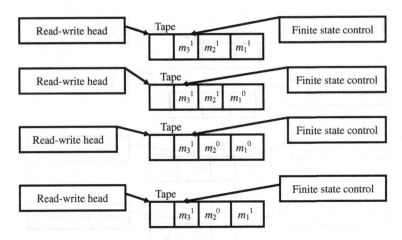

(a) The four **BMDTMs** in the second **BMPDTM**.

Fig. 5.32 Schematic representation of the current status of the execution environment to the second **BMPDTM**

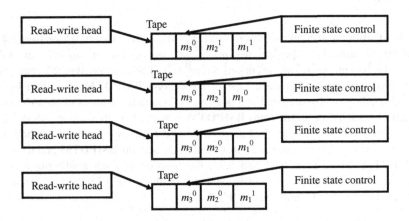

(a) The four **BMDTMs** in the third **BMPDTM**.

Fig. 5.33 Schematic representation of the current status of the execution environment to the third **BMPDTM**

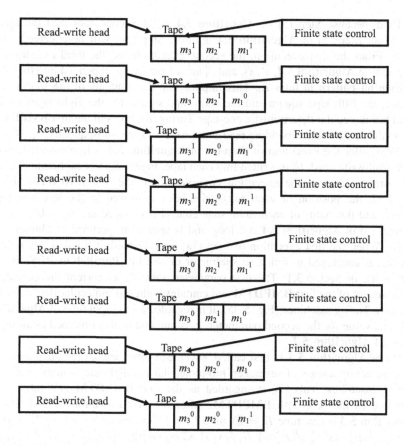

Fig. 5.34 Schematic representation of the current status of the execution environment to the first **BMPDTM**

Next, after the second execution for Step (4d) of **Algorithm 5.4** is finished, tube $T_0 = \{m_3^1\, m_2^1\, m_1^1,\ m_3^1\, m_2^1\, m_1^0,\ m_3^1\, m_2^0\, m_1^1,\ m_3^1\, m_2^0\, m_1^0,\ m_3^0\, m_2^1\, m_1^1,\ m_3^0\, m_2^1\, m_1^0,\ m_3^0\, m_2^0\, m_1^1,$ $m_3^0\, m_2^0\, m_1^0\}$, tube $T_1 = \varnothing$ and tube $T_2 = \varnothing$. This implies that the execution environment for those bio-molecular deterministic one-tape Turing machines in the second **BMPDTM** and in the third **BMPDTM** becomes the first **BMPDTM**. Figure 5.34 is employed to show the current status of the execution environment to the first **BMPDTM**. From Fig. 5.34, the contents to the eight tapes in the execution environment of the first **BMPDTM** are not changed, the position of each read-write head and the status of each finite state control are also not changed.

Then, because Step (5) of **Algorithm 5.4** is the second loop and the upper bound for Step (5) of **Algorithm 5.4** is three, Step (5a) will be executed three times. From the first execution, the second execution and the third execution of Step (5a) in **Algorithm 5.4**, s_1^1, s_2^0, and s_3^0 are, respectively, appended into the head of each bit pattern in tube T_0. This implies that the contents of the fourth tape square, the fifth tape square and the sixth tape square for the eight tapes of the eight bio-molecular deterministic one-tape Turing machines in the first **BMPDTM** are written by the corresponding read-write head and are, subsequently, s_1^1, s_2^0, and s_3^0. Next, after the execution of Step (6) in **Algorithm 5.4** is implemented, from each read-write head, b_0^0 is written into each tape. Figure 5.35 is applied to explain the current status of the execution environment to the first **BMPDTM**. From Fig. 5.35, the position of each read-write head is moved to the left new tape square, and the status of each finite state control is changed as "$b_0 = 0$".

Step (7) in **Algorithm 5.4** is a loop and is applied to perform a subtractor of n bits. When the first execution of Step (7a) is performed, it invokes **Algorithm 5.3** that is employed to finish an subtractor of one bit, **ParallelOneBitSubtractor**(T_0, k), in Sect. 5.3.1. The first parameter, tube T_0, is current the execution environment of the first **BMPDTM** and contains eight bio-molecular deterministic one-tape Turing machines (Fig. 5.35). It is regarded as an input tube of **Algorithm 5.3**. The value for the second parameter, k, is one and is also regarded as an input value of **Algorithm 5.3**.

When **Algorithm 5.3** is first called, seventeen tubes are all regarded as independent environments of seventeen bio-molecular parallel deterministic one-tape Turing machines. Tube T_0 is regarded as the first **BMPDTM** and tube T_k is regarded as the $(k + 1)$th **BMPDTM**. After the *first* execution of Step (1) in **Algorithm 5.3** is run, tube $T_0 = \varnothing$, tube $T_1 = \{b_0^0 \, m_3^0 \, m_2^0 \, m_1^1 \, m_3^1 \, m_2^1 \, m_1^1, \, b_0^0 \, s_3^0 \, s_2^0 \, s_1^1 \, m_3^1 \, m_2^0 \, m_1^1, \, b_0^0 \, s_3^0 \, s_2^0 \, s_1^1 \, m_3^1 \, m_2^1 \, m_1^1, \, b_0^0 \, s_3^0 \, s_2^0 \, s_1^1 \, m_3^0 \, m_2^0 \, m_1^1 \}$ and tube $T_2 = \{b_0^0 \, s_3^0 \, s_2^0 \, s_1^1 \, m_3^1 \, m_2^1 \, m_1^0, \, b_0^0 \, s_3^0 \, s_2^0 \, s_1^1 \, m_3^1 \, m_2^0 \, m_1^0, \, b_0^0 \, s_3^0 \, s_2^0 \, s_1^1 \, m_3^0 \, m_2^1 \, m_1^0, \, b_0^0 \, s_3^0 \, s_2^0 \, s_1^1 \, m_3^0 \, m_2^0 \, m_1^0 \}$. This implies that the new execution environments for four bio-molecular deterministic one-tape Turing machines with the content of tape square, "m_1^1", and other bio-molecular deterministic one-tape Turing machines with the content of tape square, "m_1^0" are, respectively, the second **BMPDTM** and the third **BMPDTM**. Figures 5.36 and 5.37 are applied to show the results.

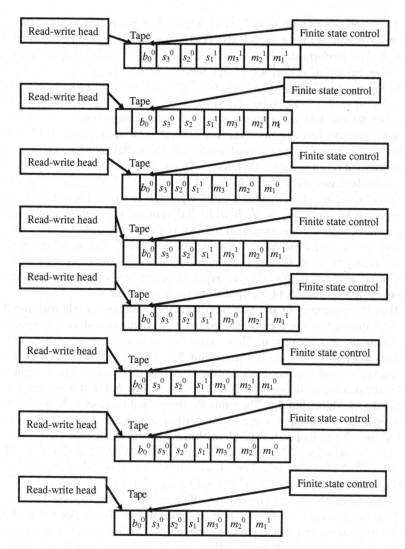

Fig. 5.35 Schematic representation of the current status of the execution environment to the first **BMPDTM**

Next, after the *first* execution of Step (2) in **Algorithm 5.3** is finished, tube $T_1 = \varnothing$, tube $T_3 = \{b_0^0 s_3^0 s_2^0 s_1^1 m_3^1 m_2^1 m_1^1, b_0^0 s_3^0 s_2^0 s_1^1 m_3^1 m_2^0 m_1^1, b_0^0 s_3^0 s_2^0 s_1^1 m_3^0 m_2^1 m_1^1, b_0^0 s_3^0 s_2^0 s_1^1 m_3^0 m_2^0 m_1^1\}$ and tube $T_4 = \varnothing$. This is to say that the new execution environment for four bio-molecular deterministic one-tape Turing machines with the content of tape square, "s_1^1" is the fourth **BMPDTM**. Figure 5.38 is applied to show the execution environment for the fourth **BMPDTM**.

Next, after the *first* execution of Step (3) in **Algorithm 5.3** is implemented, tube $T_2 = \varnothing$, tube $T_5 = \{b_0^0 s_3^0 s_2^0 s_1^1 m_3^1 m_2^1 m_1^0, b_0^0 s_3^0 s_2^0 s_1^1 m_3^1 m_2^0 m_1^0, b_0^0 s_3^0 s_2^0 s_1^1 m_3^0 m_2^1 m_1^0,$

$b_0^0 \, s_3^0 \, s_2^0 \, s_1^1 \, m_3^0 \, m_2^0 \, m_1^0\}$ and tube $T_6 = \varnothing$. Figure 5.39 is applied to show the execution environment for the sixth **BMPDTM**. After the first execution for Steps (4) through (7) is performed, tube $T_8 = \{b_0^0 \, s_3^0 \, s_2^0 \, s_1^1 \, m_3^1 \, m_2^0 \, m_1^1, \, b_0^0 \, s_3^0 \, s_2^0 \, s_1^1 \, m_3^1 \, m_2^0 \, m_1^1, \, b_0^0 \, s_3^0 \, s_2^0 \, s_1^1 \, m_3^1 \, m_2^0 \, m_1^1, \, b_0^0 \, s_3^0 \, s_2^0 \, s_1^1 \, m_3^1 \, m_2^0 \, m_1^1\}$, tube $T_{12} = \{b_0^0 \, s_3^0 \, s_2^0 \, s_1^1 \, m_3^1 \, m_2^0 \, m_1^1, \, b_0^0 \, s_3^0 \, s_2^0 \, s_1^1 \, m_3^1 \, m_2^0 \, m_1^1, \, b_0^0 \, s_3^0 \, s_2^0 \, s_1^1 \, m_3^0 \, m_2^0 \, m_1^1\}$, tube $T_7 = \varnothing$, tube $T_9 = \varnothing$, tube $T_{10} = \varnothing$, tube $T_{11} = \varnothing$, tube $T_{13} = \varnothing$, and tube $T_{14} = \varnothing$. Figures 5.40 and 5.41 are, respectively, used to show the results.

Next, after the first execution for Steps (8a) (9a) (10a) (11a) (12a) (13a) (14a) and (15a) is performed, the returned result from Steps (9a) and (13a) is "yes", and other returned result is "no". Therefore, after the first execution for Steps (9) and (13) is finished, tube $T_8 = \{b_1^0 \, d_1^0 \, b_0^0 \, s_3^0 \, s_2^0 \, s_1^1 \, m_3^1 \, m_2^1 \, m_1^1, \, b_1^0 \, d_1^0 \, b_0^0 \, s_3^0 \, s_2^0 \, s_1^1 \, m_3^1 \, m_2^0 \, m_1^1, \, b_1^0 \, d_1^0 \, b_0^0 \, s_3^0 \, s_2^0 \, s_1^1 \, m_3^1 \, m_2^0 \, m_1^1, \, b_1^0 \, d_1^0 \, b_0^0 \, s_3^0 \, s_2^0 \, s_1^1 \, m_3^1 \, m_2^0 \, m_1^1\}$, tube $T_{12} = \{b_1^1 \, d_1^1 \, b_0^0 \, s_3^0 \, s_2^0 \, s_1^1 \, m_3^1 \, m_2^1 \, m_1^0, \, b_1^1 \, d_1^1 \, b_0^0 \, s_3^0 \, s_2^0 \, s_1^1 \, m_3^1 \, m_2^0 \, m_1^0, \, b_1^1 \, d_1^1 \, b_0^0 \, s_3^0 \, s_2^0 \, s_1^1 \, m_3^1 \, m_2^0 \, m_1^0\}$. Finally, after the first execution for Step (16) is performed, tube $T_0 = \{b_1^0 \, d_1^0 \, b_0^0 \, s_3^0 \, s_2^0 \, s_1^1 \, m_3^1 \, m_2^1 \, m_1^1, \, b_1^0 \, d_1^0 \, b_0^0 \, s_3^0 \, s_2^0 \, s_1^1 \, m_3^1 \, m_2^0 \, m_1^1, \, b_1^0 \, d_1^0 \, b_0^0 \, s_3^0 \, s_2^0 \, s_1^1 \, m_3^1 \, m_2^0 \, m_1^1, \, b_1^0 \, d_1^0 \, b_0^0 \, s_3^0 \, s_2^0 \, s_1^1 \, m_3^1 \, m_2^0 \, m_1^1, \, b_1^1 \, d_1^1 \, b_0^0 \, s_3^0 \, s_2^0 \, s_1^1 \, m_3^1 \, m_2^0 \, m_1^0\}$ and other tubes become all empty tubes. Figures 5.42, 5.43 and 5.44 are applied to show the result.

After the execution of the *first* time for each operation in **Algorithm 5.3** is finished, the subtraction for the first bit for each minuend and its corresponding subtrahend is also performed. Then, when the *second* execution of Step (7a) is implemented, it again invokes **Algorithm 5.3**. The first parameter, tube T_0, is current the execution environment of the first **BMPDTM** and contains eight biomolecular deterministic one-tape Turing machines (Fig. 5.44). It is regarded as an input tube of **Algorithm 5.3**. The value for the second parameter, k, is *two* and is also regarded as an input value of **Algorithm 5.3**. After the execution of the *second* time for each operation in **Algorithm 5.3** is performed, tube $T_0 = \{b_2^0 \, d_2^1 \, b_1^0 \, d_1^0 \, b_0^0 \, s_3^0 \, s_2^0 \, s_1^1 \, m_3^1 \, m_2^1 \, m_1^1, \, b_2^0 \, d_2^1 \, b_1^0 \, d_1^0 \, b_0^0 \, s_3^0 \, s_2^0 \, s_1^1 \, m_3^1 \, m_2^0 \, m_1^1, \, b_2^0 \, d_2^1 \, b_1^0 \, d_1^0 \, b_0^0 \, s_3^0 \, s_2^0 \, s_1^1 \, m_3^1 \, m_2^0 \, m_1^1, \, b_2^0 \, d_2^0 \, b_1^0 \, d_1^0 \, b_0^0 \, s_3^0 \, s_2^0 \, s_1^1 \, m_3^1 \, m_2^0 \, m_1^1, \, b_2^1 \, d_2^1 \, b_1^0 \, d_1^0 \, b_0^0 \, s_3^0 \, s_2^0 \, s_1^1 \, m_3^1 \, m_2^0 \, m_1^0\}$ and other tubes become all empty tubes. Figure 5.45 is used to show the result.

After the execution of the *second* time for each operation in **Algorithm 5.3** is performed, the subtraction for the *second* bit for each minuend and its corresponding subtrahend is also finished. Next, when the *third* execution of Step (7a) is implemented, it again invokes **Algorithm 5.3**. The first parameter, tube T_0, is current the execution environment of the first **BMPDTM** and contains eight biomolecular deterministic one-tape Turing machines (Fig. 5.45). It is regarded as an input tube of **Algorithm 5.3**. The value for the second parameter, k, is *three* and is also regarded as an input value of **Algorithm 5.3**. After the execution of the *third* time for each operation in **Algorithm 5.3** is performed, tube $T_0 = \{b_3^0 \, d_3^1 \, b_2^0 \, d_2^1 \, b_1^0 \, d_1^0 \, b_0^0 \, s_3^0 \, s_2^0 \, s_1^1 \, m_3^1 \, m_2^1 \, m_1^1, \, b_3^0 \, d_3^1 \, b_2^0 \, d_2^1 \, b_1^0 \, d_1^0 \, b_0^0 \, s_3^0 \, s_2^0 \, s_1^1 \, m_3^1 \, m_2^0 \, m_1^1, \, b_3^0 \, d_3^0 \, b_2^0 \, d_2^0 \, b_1^0 \, d_1^0 \, b_0^0 \, s_3^0 \, s_2^0 \, s_1^1 \, m_3^1 \, m_2^0 \, m_1^1, \, b_3^0 \, d_3^1 \, b_2^0 \, d_2^0 \, b_1^0 \, d_1^0 \, b_0^0 \, s_3^0 \, s_2^0 \, s_1^1 \, m_3^1 \, m_2^0 \, m_1^1, \, b_3^1 \, d_3^1 \, b_2^1 \, d_2^1 \, b_1^1 \, d_1^1 \, b_0^0 \, s_3^0 \, s_2^0 \, s_1^1 \, m_3^0 \, m_2^0 \, m_1^0\}$ and other tubes become all empty tubes. Figure 5.46 is used to show the result and **Algorithm 5.4** is terminated.

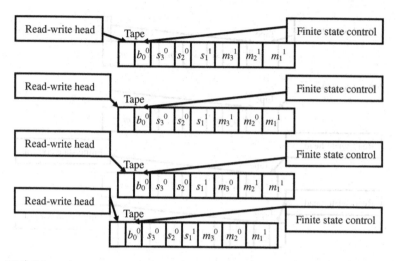

Fig. 5.36 Schematic representation of the current status of the execution environment to the second **BMPDTM**

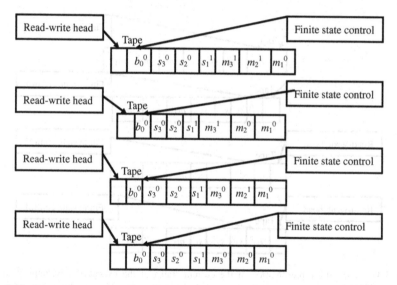

Fig. 5.37 Schematic representation of the current status of the execution environment to the third **BMPDTM**

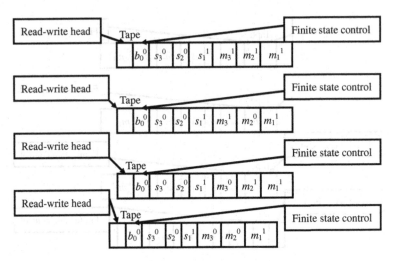

Fig. 5.38 Schematic representation of the current status of the execution environment to the fourth **BMPDTM**

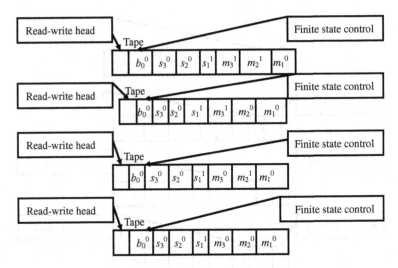

Fig. 5.39 Schematic representation of the current status of the execution environment to the sixth **BMPDTM**

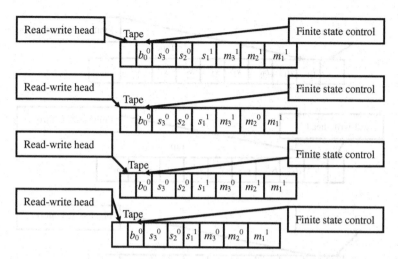

Fig. 5.40 Schematic representation of the current status of the execution environment to the ninth **BMPDTM**

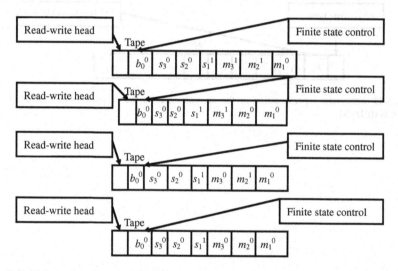

Fig. 5.41 Schematic representation of the current status of the execution environment to the 13th **BMPDTM**

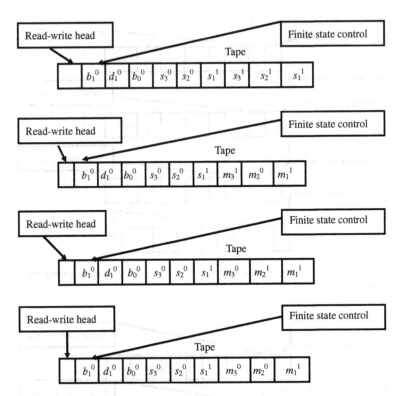

Fig. 5.42 Schematic representation of the current status of the execution environment to the ninth **BMPDTM**

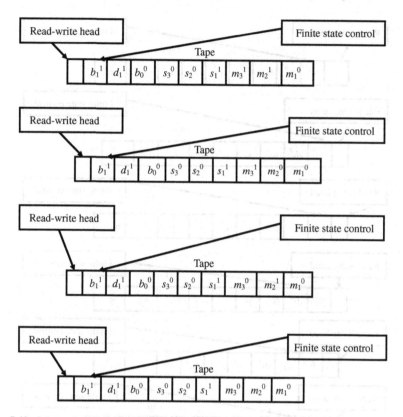

Fig. 5.43 Schematic representation of the current status of the execution environment to the 13th **BMPDTM**

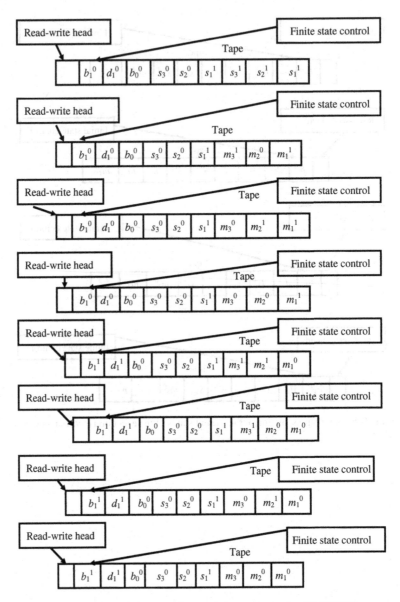

Fig. 5.44 The current status of the execution environment to the first **BMPDTM**

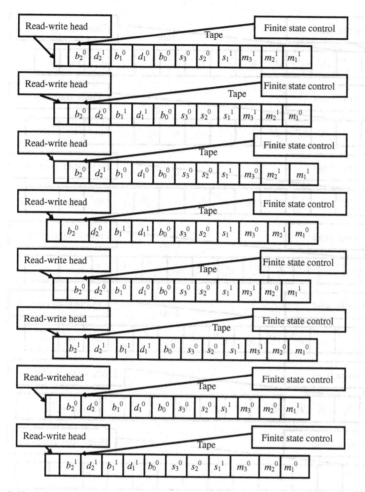

Fig. 5.45 Schematic representation of the current status of the execution environment to the first **BMPDTM**

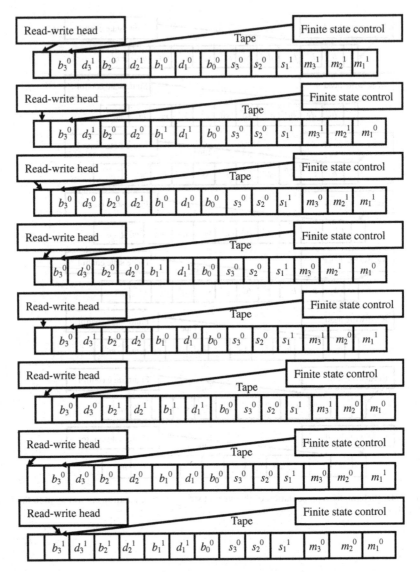

Fig. 5.46 Schematic representation of the current status of the execution environment to the first **BMPDTM**

5.4 Summary

In this chapter an introduction to how arithmetic operations on bits were implemented by means of bio-molecular operations was provided. We introduced that the *bio-molecular parallel deterministic one-tape Turing machine* (abbreviated **BMPDTM**) was pictured schematically in Fig. 5.2 and this machine includes

n bio-molecular *deterministic one-tape Turing machines*. We then described that each bio-molecular *deterministic one-tape Turing machine* (abbreviated **BMDTM**) contained a finite state control, a read-write head, and a tape made up of a two-way infinite sequence of tape squares (shown in Fig. 5.2). We also introduced that each bit pattern encoded in a tube in a bio-molecular computer was regarded as a tape in a bio-molecular deterministic one-tape Turing machine and that for each tape in a **BMPDTM**, each biological operation was applied to simultaneously perform the function of the corresponding finite state control and read-write head.

We then described **Algorithm 5.1** and its proof to show how the Boolean function of a parallel one-bit adder was implemented by means of biological operations. We also introduced **Algorithm 5.2** and its proof to explain how the arithmetic sum for a parallel adder of n-bits was computed by means of molecular operations. We then gave one example to reveal the power of a parallel adder of n-bits. We also introduced **Algorithm 5.3** and its proof to demonstrate how the Boolean function of a parallel one-bit subtractor was implemented by means of biological operations. We also described **Algorithm 5.4** and its proof to show how the arithmetic difference for a parallel subtractor of n-bits was implemented by means of molecular operations. We then gave one example to explain the power of a parallel subtractor of n-bits.

5.5 Bibliographical Notes

The textbook by Hopcroft et al. (Hopcroft et al. 2006) is a good introduction to automata theory. The book by Hofstadter (Hofstadter 1980) is a good introduction to theory of computation on discussing the Godel number and its application. The books that were written by Hennie (Hennie 1977), Kfoury (Kfoury 1982), Minsky (Minsky 1967) and Sipser (Sipser 2005) are good introduction to the complexity of problems. The books which were written by the authors in Brown and Vranesic (2007), Mano (1979, 1993), Null and Lobur (2010), Shiva (2008), Stalling (2000) are good introduction to digital arithmetic and logical circuits.

5.6 Exercises

5.1 The binary operator $\vee$ defines logical operation **OR**. The truth table to a logical operation, $x \vee 0$, is shown in Table 5.3, where x is a Boolean variable that is the first input and 0 is the second input. Based on Table 5.3, write a bio-molecular program to implement the function of the logical operation, $x \vee 0$.

Table 5.3 The truth table to a logical operation, $x \vee 0$, is shown

The first input (x)	The second input	$x \vee 0$
0	0	0
1	0	1

5.2 The binary operator $\wedge$ defines logical operation **AND**. The truth table to a logical operation, $x \wedge 1$, is shown in Table 5.4, where x is a Boolean variable that is the first input and 1 is the second input. Based on Table 5.4, write a bio-molecular program to implement the function of the logical operation, $x \wedge 1$.

Table 5.4 The truth table to a logical operation, $x \wedge 1$, is shown

The first input (x)	The second input	$x \wedge 1$
0	1	0
1	1	1

5.3 The binary operator $\vee$ defines logical operation **OR**, and the unary operator $'$ defines logical operation **NOT**. The truth table to a logical operation, $x \vee x'$, is shown in Table 5.5, where x is a Boolean variable that is the first input and x' is its negation that is the second input. Based on Table 5.5, write a bio-molecular program to implement the function of the logical operation, $x \vee x'$.

Table 5.5 The truth table to a logical operation, $x \vee x'$, is shown

The first input (x)	The second input (x')	$x \vee x'$
0	1	1
1	0	1

5.4 The binary operator $\wedge$ defines logical operation **AND**, and the unary operator $'$ defines logical operation **NOT**. The truth table to a logical operation, $x \wedge x'$, is shown in Table 5.6, where x is a Boolean variable that is the first input and x' is its negation that is the second input. Based on Table 5.6, write a bio-molecular program to implement the function of the logical operation, $x \wedge x'$.

Table 5.6 The truth table to a logical operation, $x \wedge x'$, is shown

The first input (x)	The second input (x')	$x \wedge x'$
0	1	0
1	0	0

5.5 The binary operator $\vee$ defines logical operation **OR**. The truth table to a logical operation, $x \vee x$, is shown in Table 5.7, where x is a Boolean variable that is the first input and the second input. Based on Table 5.7, write a bio-molecular program to implement the function of the logical operation, $x \vee x$.

Table 5.7 The truth table to a logical operation, $x \vee x$, is shown

The first input (x)	The second input (x)	$x \vee x$
0	0	0
1	1	1

5.6 The binary operator $\wedge$ defines logical operation **AND**. The truth table to a logical operation, $x \wedge x$, is shown in Table 5.8, where x is a Boolean variable that is the first input and the second input. Based on Table 5.8, write a bio-molecular program to implement the function of the logical operation, $x \wedge x$.

Table 5.8 The truth table to a logical operation, $x \wedge x$, is shown

The first input (x)	The second input (x)	$x \wedge x$
0	0	0
1	1	1

5.7 The binary operator $\vee$ defines logical operation **OR**. The truth table to a logical operation, $x \vee 1$, is shown in Table 5.9, where x is a Boolean variable that is the first input and 1 is the second input. Based on Table 5.9, write a bio-molecular program to implement the function of the logical operation, $x \vee 1$.

Table 5.9 The truth table to a logical operation, $x \vee 1$, is shown

The first input (x)	The second input	$x \vee 1$
0	1	1
1	1	1

5.8 The binary operator $\wedge$ defines logical operation **AND**. The truth table to a logical operation, $x \wedge 0$, is shown in Table 5.10, where x is a Boolean variable that is the first input and 0 is the second input. Based on Table 5.10, write a bio-molecular program to implement the function of the logical operation, $x \wedge 0$.

Table 5.10 The truth table to a logical operation, $x \wedge 0$, is shown

The first input (x)	The second input	$x \wedge 0$
0	0	0
1	0	0

5.9 The unary operator $'$ defines logical operation **NOT**. The truth table to a logical operation $(x')'$, is shown in Table 5.11, where x is a Boolean variable that is the first input. Based on Table 5.11, write a bio-molecular program to implement the function of the logical operation $(x')'$.

Table 5.11 The truth table to a logical operation $(x')'$, is shown

The first input (x)	x'	$(x')'$
0	1	0
1	0	1

5.10 The binary operator $\wedge$ defines logical operation **AND**, and the unary operator $'$ defines logical operation **NOT**. The truth table to a logical operation, $x \wedge y'$, is shown in Table 5.12, where x and y are two Boolean variables that are respectively the first input and the second input. Based on Table 5.12, write a bio-molecular program to implement the function of the logical operation, $x \wedge y'$.

Table 5.12 The truth table to a logical operation, $x \wedge y'$, is shown

The first input (x)	The second input (y)	$x \wedge y'$
0	0	0
0	1	0
1	0	1
1	1	0

References

W.-L. Chang, M. Ho, M. Guo, Fast parallel molecular algorithms for DNA-based computation: factoring integers. IEEE Trans. Nanobiosci. **4**(2), 149–163 (2005)

M. Ho, Fast parallel molecular solutions for DNA-based supercomputing: the subset-product problem. Biosystems **80**(3), 233–250 (2005)

J. Hopcroft, R. Motwani, J. Ullman, *Introduction to Automata Theory, Languages, and Computation* (Addison Wesley, Boston, 2006), ISBN: 81-7808-347-7

F. Hennie, *Introduction to Computability* (Addison Wesley, Boston, 1977), ISBN: 0201028484

A. Kfoury, R. Moll, A. Michael, *A Programming Approach to Computability* (Springer, Berlin, 1982), ISBN: 0387907432

M. Minsky, *Computation: Finite and Infinite Machines* (Prentice-Hall, Englewood Cliffs, 1967) ISBN: 0-13-165563-9

M. Sipser, *Introduction the Theory of Computation* (Course Technology, Boston, 2005), ISBN: 0534950973

S. Brown, Z. Vranesic, *Fundamentals of Digital Logic with Verilog Design* (McGraw-Hill, New York, 2007), ISBN: 978-0077211646

M.M. Mano, *Digital Logic and Computer Design* (Prentice-Hall, Englewood Cliffs, 1979), ISBN: 0-13-214510-3

M.M. Mano, *Computer System Architecture* (Prentice Hall, Englewood Cliffs, 1993), ISBN: 978-0131755635

L. Null, J. Lobur, *Essentials of Computer Organization and Architecture* (Jones and Bartlett Learning, Sudbury, 2010), ISBN: 978-1449600068

S.G. Shiva, *Computer Organization, Design, and Architecture* (CRC Press, Boca Raton, 2008), ISBN: 9780849304163

W. Stalling, *Computer Organization and Architecture* (Prentice Hall, Englewood Cliffs, 2000), ISBN: 978-0132936330

D. Hofstadter, *Godel, Escher, Bach: An Eternal Golden Braid* (Vintage, St. Paul, 1980), ISBN: 039474502

Chapter 6
Introduction to Logic Operations on Bits on Bio-molecular Computer

In Chap. 5, we showed how to perform arithmetic operations on bits in a bio-molecular computer. In this chapter, we show how to finish logic operations on bits in a bio-molecular computer. Those logic operations on bits are **NOT**, **OR**, **AND**, **NOR**, **NAND**, **Exclusive-OR (XOR)** and **Exclusive-NOR (XNOR)**, and are shown in Fig. 6.1. The *bio-molecular parallel deterministic one-tape Turing machine* (abbreviated **BMPDTM**) denoted in Chap. 5 is chosen as our model for the purpose of clearly explaining how those logic operations on bits are performed.

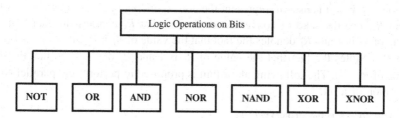

Fig. 6.1 Logic operations on bits

6.1 Introduction to NOT Operation on Bio-molecular Computer

The **NOT** operation of a bit inverts or provides the one's complement for the bit. This operation takes a single input and generates one output. The **NOT** operation of a bit offers the following result:

$$\textbf{NOT}\,1 = 0$$
$$\textbf{NOT}\,0 = 1$$

Hence, **NOT** of a Boolean variable R, written as $\bar{R}$ is 1 if and only if R is 0. Similarly, $\bar{R}$ is 0 if and only if R is 1. This definition may also be specified in the form of a truth table:

W.-L. Chang and A. V. Vasilakos, *Molecular Computing*, Studies in Big Data 4, DOI: 10.1007/978-3-319-05122-2_6, © Springer International Publishing Switzerland 2014

Table 6.1 The truth table for the **NOT** operation of a bit for a Boolean variable R

Input R	Output $\bar{R}$
0	1
1	0

The **NOT** operation of a bit inverts the one's complement for the only input. The **NOT** operation of n-bits provides the corresponding one's complement for each input in n inputs by means of performing the **NOT** operation of a bit of n times. The following subsections will be applied to describe how to finish the **NOT** operation of a bit and the **NOT** operation of n-bits.

6.1.1 The Construction for the Parallel NOT Operation of a Bit on Bio-molecular Computer

Suppose that a one-bit binary number, R_k, is applied to represent the input for the **NOT** operation of a bit for $1 \leq k \leq n$. Also suppose that a one-bit binary number $\bar{R}_k$ for $1 \leq k \leq n$ is used to represent the corresponding one's complement for the input R_k. For the sake of convenience, assume that R_k^1 denotes the fact that the value of R_k is 1 and R_k^0 denotes the fact that the value of R_k is 0. Similarly, suppose that $\bar{R}_k^1$ denotes the fact that the value of $\bar{R}_k$ is 1 and $\bar{R}_k^0$ denotes the fact that the value of $\bar{R}_k$ is 0. The following algorithm is proposed to perform the parallel NOT operation of a bit.

> **Algorithm 6.1: ParallelNOT(T_0, k)**
>
> (1) $T_1 = +(T_0, R_k^1)$ and $T_2 = -(T_0, R_k^1)$.
>
> (2a) **If** (Detect(T_1) $==$ "yes") **then**
>
> (2) Append-head(T_1, $\bar{R}_k^0$).
>
> **EndIf**
>
> (3a) **If** (Detect(T_2) $==$ "yes") **then**
>
> (3) Append-head(T_2, $\bar{R}_k^1$).
>
> **EndIf**
>
> (4) $T_0 = \cup(T_1, T_2)$.
>
> **EndAlgorithm**

Lemma 6-1: *The algorithm,* **ParallelNOT(T_0, k),** *parallel* **NOT** *operation of a bit.*

Proof The algorithm, **ParallelNOT**(T_0, k), is implemented by means of the *extract, detect, append-head* and *merge* operations. Step (1) uses the *extract* operations to form some different tubes (T_1 and T_2) including different inputs. That is, T_1 includes all of the inputs that have $R_k = 1$ and T_2 consists of all of the inputs that have $R_k = 0$. Having performed Step (1), this implies that two different inputs for the **NOT** operation of a bit as shown in Table 6.1 were poured into tubes T_1 through T_2, respectively.

Steps (2a) and (3a) are, respectively, used to test whether contains any input for tubes T_1 and T_2 or not. If any a "yes" is returned for those steps, then the corresponding *append-head* operations will be run. Next, Steps (2) through (3) use the *append-head* operations to append $\bar{R}_k^0$ and $\bar{R}_k^1$ onto the head of every input in the corresponding tubes. After performing Steps (2) through (3), we can say that two different outputs to the **NOT** operation of a bit as shown in Table 6.1 are appended into tubes T_1 through T_2. Finally, the execution of Step (4) applies the *merge* operation to pour tubes T_1 through T_2 into tube T_0. Tube T_0 contains the result performing the NOT operation of a bit as shown in Table 6.1. ∎

6.1.2 The Construction for the Parallel NOT Operation of N Bits on Bio-molecular Computer

The parallel **NOT** operation of n bits simultaneously inverts the corresponding one's complement for 2^n combinations of n bits. The following algorithm is offered to perform the parallel **NOT** operation of n bits. Notations in **Algorithm 6.2** are denoted in Sect. 6.1.1.

Algorithm 6.2: N-Bits-ParallelNOT(T_0)

(1) Append-head(T_1, R_1^1).

(2) Append-head(T_2, R_1^0).

(3) $T_0 = \cup(T_1, T_2)$.

(4) **For** $k = 2$ **to** n

      (4a) Amplify(T_0, T_1, T_2).

      (4b) Append-head(T_1, R_k^1).

      (4c) Append-head(T_2, R_k^0).

      (4d) $T_0 = \cup(T_1, T_2)$.

EndFor

(5) **For** $k = 1$ **to** n

     (5a) **ParallelNOT**(T_0, k).

EndFor

EndAlgorithm

Lemma 6-2: *The algorithm,* **N-Bits-ParallelNOT**(T_0), *can be applied to perform the parallel* **NOT** *operation of n bits.*

Proof Steps (1) through (4d) are mainly applied to construct solution space of 2^n unsigned integers (the range of values for them is from 0 to $2^n - 1$). After they are performed, tube T_0 includes those inputs encoding 2^n unsigned integers. Next, Step (5) is the main loop and is mainly used to perform the parallel **NOT** operation of n bits. Each execution of Step (5a) calls **ParallelNOT**(T_0, k) in Sect. 6.1.1 to invert the one's complement for the kth bit of each input in 2^n inputs. Repeat execution of Step (5a) until the nth bit of each input in 2^n inputs is processed. Tube T_0 contains the result performing the parallel **NOT** operation of n bits. ∎

6.1.3 The Power for the Parallel NOT Operation of N Bits on Bio-molecular Computer

Consider that four values for an unsigned integer of two bits are, subsequently, $00(0_{10})$ $\left(R_2^0 R_1^0\right)$, $01(1_{10})$ $\left(R_2^0 R_1^1\right)$, $10(2_{10})$ $\left(R_2^1 R_1^0\right)$ and $11(3_{10})$ $\left(R_2^1 R_1^1\right)$. We want to simultaneously invert the corresponding one's complement for those four values. **Algorithm 6.2, N-Bits-ParallelNOT**(T_0), can be applied to run the task. Tube T_0 is an empty tube and is regarded as an input tube of **Algorithm 6.2**. In light of Definition 5–2, the input tube T_0 is regarded as the execution environment of the first **BMPDTM**. Similarly, tubes T_1 and T_2 used in **Algorithm 6.2** also are regarded, subsequently, as the execution environment of the second **BMPDTM** and the execution environment of the third **BMPDTM**.

Steps (1) through (4d) in **Algorithm 6.2** are applied to generate a **BMPDTM** with four bio-molecular deterministic one-tape Turing machines. After the execution for Step (1) and Step (2) of **Algorithm 6.2** is implemented, tube $T_1 = \left\{R_1^1\right\}$ and tube $T_2 = \left\{R_1^0\right\}$. This is to say that a **BMDTM** in the second **BMPDTM** and in the third **BMPDTM** is generated. Figure 6.2 is used to explain the current status of the execution environment to the second **BMPDTM** and the third **BMPDTM**. From Fig. 6.2, the content of the first tape square for the tape in the first **BMDTM** in the second **BMPDTM** is written by its corresponding read-write head and is 1 ($R_1 = 1$), and the content of the first tape square for the tape in the first **BMDTM** in the third **BMPDTM** is written by its corresponding read-write head and is 0 ($R_1 = 0$). For the first **BMDTM** in the second **BMPDTM**, the position of the read-write head is moved to the *left new* tape square, and the state of the finite state control is changed as "$R_1 = 1$". Similarly, for the first **BMDTM** in the third **BMPDTM**, the position of the read-write head is moved to the *left new* tape square, and the state of the finite state control is changed as "$R_1 = 0$".

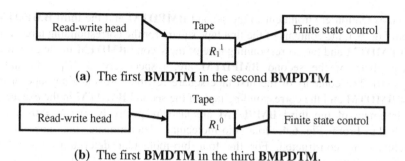

(a) The first **BMDTM** in the second **BMPDTM**.

(b) The first **BMDTM** in the third **BMPDTM**.

Fig. 6.2 Schematic representation of the current status of the execution environment to the second **BMPDTM** and the third **BMPDTM**

Next, after the execution for Step (3) of **Algorithm 6.2** is performed, tube $T_0 = \{R_1^1, R_1^0\}$, tube $T_1 = \varnothing$ and tube $T_2 = \varnothing$. This is to say that the execution environment for the first bio-molecular deterministic one-tape Turing machine in the second **BMPDTM** and the first bio-molecular deterministic one-tape Turing machine in the third **BMPDTM** becomes the first **BMPDTM**. The position of the corresponding read-write head and the state of the corresponding finite state control are both reserved. Figure 6.3 is applied to illustrate the current status of the execution environment to the first **BMPDTM**. From Fig. 6.3, the contents to the two tapes in the execution environment of the first **BMPDTM** are not changed.

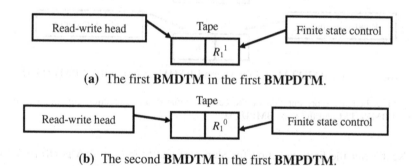

(a) The first **BMDTM** in the first **BMPDTM**.

(b) The second **BMDTM** in the first **BMPDTM**.

Fig. 6.3 Schematic representation of the current status of the execution environment to the first **BMPDTM**

Step (4) is the first loop in **Algorithm 6.2**, the upper bound (n) is two because the number of bits for representing those four values is two. Hence, after the first execution of Step (4a) is finished, tube $T_0 = \varnothing$, tube $T_1 = \{R_1^1, R_1^0\}$ and tube $T_2 = \{R_1^1, R_1^0\}$. This implies that the first **BMDTM** and the second **BMDTM** in the execution environment of the first **BMPDTM** are both copied into the second **BMPDTM** and the third **BMPDTM**. Figure 6.4 is used to show the current status of the execution environment to the second **BMPDTM** and the third **BMPDTM**. From Fig. 6.4, the contents of the first tape square for the corresponding tape of the first **BMDTM** and the corresponding tape of the second **BMDTM** in the execution environment of the second **BMPDTM** are, respectively, 1 ($R_1 = 1$) and 0 ($R_1 = 0$). The contents of the first tape square for the corresponding tape of the first **BMDTM** and the corresponding tape of the second **BMDTM** in the execution environment of the third **BMPDTM** are also, respectively, 0 ($R_1 = 0$) and 1 ($R_1 = 1$). From Fig. 6.4, four bio-molecular deterministic one-tape Turing machines are constructed. For the four bio-molecular deterministic one-tape Turing machines, the position of the corresponding read-write head and the state of the corresponding finite state control are reserved.

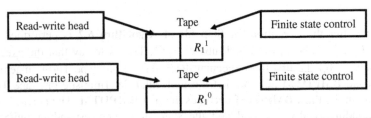

(a) The first **BMDTM** and the second **BMDTM** in the second **BMPDTM**.

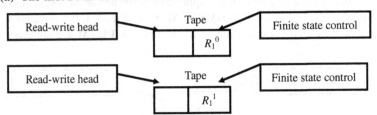

(b) The first **BMDTM** and the second **BMDTM** in the third **BMPDTM**.

Fig. 6.4 Schematic representation of the current status of the execution environment to the second **BMPDTM** and the third **BMPDTM**

Next, after the first execution for Step (4b) and Step (4c) of **Algorithm 6.2** is implemented, tube $T_1 = \{R_2^1 R_1^1, R_2^1 R_1^0\}$ and tube $T_2 = \{R_2^0 R_1^1, R_2^0 R_1^0\}$. This indicates that the content of the *second* tape square for the tape in the first **BMDTM** in the second **BMPDTM** is written by its corresponding read-write head and is 1 ($R_2 = 1$), and the content of the *second* tape square for the tape in the second **BMDTM** in the second **BMPDTM** is written by its corresponding read-write head and is also 1 ($R_2 = 1$). Simultaneously, the position of the corresponding read-write head is both moved to the *left new* tape square and the state of the corresponding finite state control is both changed as "$R_2 = 1$". Similarly, the content of the *second* tape square for the tape in the first **BMDTM** in the third **BMPDTM** is written by its corresponding read-write head and is 0 ($R_2 = 0$), and the content of the *second* tape square for the tape in the second **BMDTM** in the third **BMPDTM** is written by its corresponding read-write head and is also 0 ($R_2 = 0$). Simultaneously, the position of the corresponding read-write head is both moved to the *left new* tape square and the state of the corresponding finite state control is both changed as "$R_2 = 0$". Figure 6.5 is applied to explain the current status of the execution environment to the second **BMPDTM** and the third **BMPDTM**.

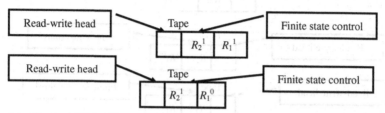

(a) The first **BMDTM** and the second **BMDTM** in the second **BMPDTM**.

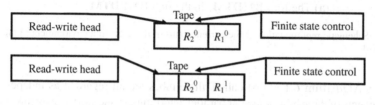

(b) The first **BMDTM** and the second **BMDTM** in the third **BMPDTM**.

Fig. 6.5 Schematic representation of the current status of the execution environment to the second **BMPDTM** and the third **BMPDTM**

Next, after the first execution for Step (4d) of **Algorithm 6.2** is performed, tube $T_0 = \{R_2^1 R_1^1, R_2^1 R_1^0, R_2^0 R_1^1, R_2^0 R_1^0\}$, tube $T_1 = \varnothing$ and tube $T_2 = \varnothing$. This implies that the execution environment for those bio-molecular deterministic one-tape Turing machines in the second **BMPDTM** and in the third **BMPDTM** becomes the first **BMPDTM**. Figure 6.6 is employed to illustrate the current status of the execution environment to the first **BMPDTM**. From Fig. 6.6, the contents to the four tapes in the execution environment of the first **BMPDTM** are not changed, and the position of the corresponding read-write head and the state of the corresponding finite state control are reserved.

Step (5) in **Algorithm 6.2** is a loop and is used to perform the parallel **NOT** operation of n bits. When the first execution of Step (5a) is implemented, it invokes **Algorithm 6.1** that is employed to finish the parallel **NOT** operation of one bit, **ParallelNOT**(T_0, k), in Sect. 6.1.1. The first parameter, tube T_0, is current the execution environment of the first **BMPDTM** and contains four bio-molecular deterministic one-tape Turing machines (Fig. 6.6). It is regarded as an input tube of **Algorithm 6.1**. The value for the second parameter, k, is one and is also regarded as an input value of **Algorithm 6.1**.

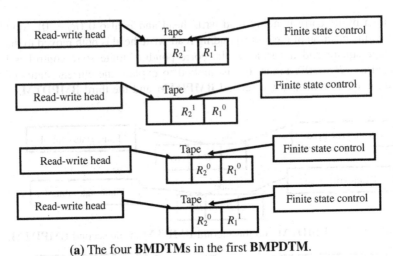

(a) The four **BMDTMs** in the first **BMPDTM**.

Fig. 6.6 Schematic representation of the current status of the execution environment to the first **BMPDTM**

When **Algorithm 6.1** is first called, three tubes are all regarded as independent environments of three bio-molecular parallel deterministic one-tape Turing machines. Tube T_0 is regarded as the first **BMPDTM** and tube T_k is regarded as the $(k + 1)$th **BMPDTM**. After the *first* execution of Step (1) in **Algorithm 6.1** is implemented, tube $T_0 = \varnothing$, tube $T_1 = \{R_2^1 R_1^1, R_2^0 R_1^1\}$ and tube $T_2 = \{R_2^1 R_1^0, R_2^0 R_1^0\}$. This indicates that the new execution environments for two bio-molecular deterministic one-tape Turing machines with the content of tape square, "R_1^1", and other two bio-smolecular deterministic one-tape Turing machines with the content of tape square, "R_1^0" are, respectively, the second **BMPDTM** and the third **BMPDTM**. The position of the corresponding read-write head and the state of the corresponding finite state control are reserved. Figures 6.7 and 6.8 are applied to show the results.

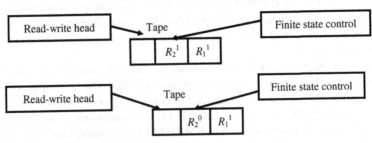

(a) The two **BMDTMs** in the second **BMPDTM**.

Fig. 6.7 Schematic representation of the current status of the execution environment to the second **BMPDTM**

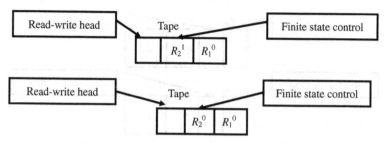

(a) The two **BMDTMs** in the third **BMPDTM**.

Fig. 6.8 Schematic representation of the current status of the execution environment to the third **BMPDTM**

Next, after the first execution for Steps (2a) and (3a) is implemented, the returned result from Steps (2a) and (3a) is "yes". Therefore, after the first execution for Steps (2) and (3) is performed, tube $T_1 = \{\bar{R}_1^0 R_2^1 R_1^1, \ \bar{R}_1^0 R_2^0 R_1^1\}$ and tube $T_2 = \{\bar{R}_1^1 R_2^1 R_1^0, \ \bar{R}_1^1 R_2^0 R_1^0\}$. Figures 6.9 and 6.10 are applied to respectively illustrate the current status of the execution environment to the second **BMDTM** and the third **BMPDTM**. From Figs. 6.9 and 6.10, the content of the third tape square for the tape in the first **BMDTM** and the second **BMDTM** in the second **BMPDTM** is both written by the corresponding read-write head and is 0 ($\bar{R}_1 = 0$), and the content of the third tape square for the tape in the first **BMDTM** and the second **BMDTM** in the third **BMPDTM** is both written by the corresponding read-write head and is 1 ($\bar{R}_1 = 1$).

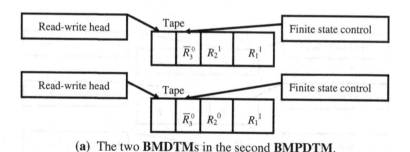

(a) The two **BMDTMs** in the second **BMPDTM**.

Fig. 6.9 Schematic representation of the current status of the execution environment to the second **BMPDTM**

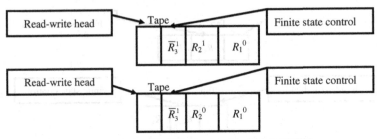

(a) The two **BMDTMs** in the third **BMPDTM**.

Fig. 6.10 Schematic representation of the current status of the execution environment to the third **BMPDTM**

For the two **BMDTMs** in the second **BMPDTM**, the position of the corresponding read-write head is both moved to the *left new* tape square, and the state of the corresponding finite state control is both changed as "$\bar{R}_3 = 0$". Similarly, for the other two **BMDTMs** in the third **BMPDTM**, the position of the corresponding read-write head is both moved to the *left new* tape square, and the state of the corresponding finite state control is both changed as "$\bar{R}_3 = 1$". Finally, after the first execution for Step (4) is finished, tube $T_1 = \{\bar{R}_1^0 R_2^1 R_1^1, \ \bar{R}_1^0 R_2^0 R_1^1, \ \bar{R}_1^1 R_2^1 R_1^0, \ \bar{R}_1^1 R_2^0 R_1^0\}$, tube $T_1 = \varnothing$ and tube $T_2 = \varnothing$. This implies that the execution environment for the first and second bio-molecular deterministic one-tape Turing machines in the second **BMPDTM** and the first and second bio-molecular deterministic one-tape Turing machines in the third **BMPDTM** becomes the first **BMPDTM**. The position of the corresponding read-write head and the state of the corresponding finite state control are all reserved. Figure 6.11 is applied to show the result.

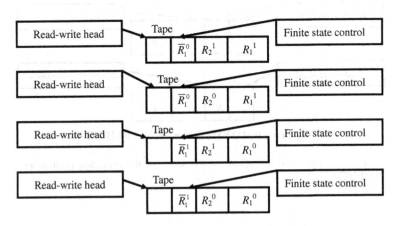

(a) The four **BMDTMs** in the first **BMPDTM**.

Fig. 6.11 Schematic representation of the current status of the execution environment to the first **BMPDTM**

After the execution of the *first* time for each operation in **Algorithm 6.**1 is performed, the parallel **NOT** operation to the first bit of the four inputs is also finished. Then, when the *second* execution of Step (5a) in **Algorithm 6.2** is implemented, it again invokes **Algorithm 6.1**. The first parameter, tube T_0, is current the execution environment of the first **BMPDTM** and contains four bio-molecular deterministic one-tape Turing machines (Fig. 6.11). It is regarded as an input tube of **Algorithm 6.1**. The value for the second parameter, k, is *two* and is also regarded as an input value of **Algorithm 6.1**. After the execution of the *second* time for each operation in **Algorithm 6.1** is performed, tube $T_0 = \{\bar{R}_2^0 \bar{R}_1^0 R_2^1 R_1^1, \ \bar{R}_2^1 \bar{R}_1^0 R_2^0 R_1^1, \ R_2^1 R_1^0, \ \bar{R}_2^0 \bar{R}_1^1 \bar{R}_2^1 \bar{R}_1^1 R_2^0 R_1^0\}$ and other tubes become all empty tubes. This indicates that the content of the fourth tape square for the four **BMDTMs** is written by the corresponding read-write head and is, respectively, $\bar{R}_2^0$, $\bar{R}_2^1$, $\bar{R}_2^0$ and $\bar{R}_2^1$. Figure 6.12 is used to explain the result and **Algorithm 6.2** is terminated.

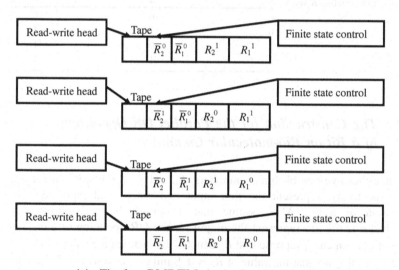

(a) The four **BMDTMs** in the first **BMPDTM**.

Fig. 6.12 Schematic representation of the current status of the execution environment to the first **BMPDTM**

6.2 Introduction to OR Operation on Bio-molecular Computer

The **OR** operation of a bit for two Boolean variables R and S produces a result of 1 if R or S, or both, are 1. However, if both R and S are zero, then the result is 0. A plus sign $+$ (logical sum) or $\vee$ symbol is normally applied to represent **OR**.

The four possible combinations for the OR operation of a bit to two Boolean variables R and S are

$$0 \vee 0 = 0$$
$$0 \vee 1 = 1$$
$$1 \vee 0 = 1$$
$$1 \vee 1 = 1$$

A truth table is usually employed with logic operation to represent all possible combinations of inputs and the corresponding outputs. The truth table for the OR operation is shown in Table 6.2.

Table 6.2 The truth table for the OR operation of a bit for two Boolean variables R and S	Input		Output
	R	S	$C = R \vee S$
	0	0	0
	0	1	1
	1	0	1
	1	1	1

6.2.1 The Construction for the Parallel OR Operation of a Bit on Bio-molecular Computer

Assume that two one-bit binary numbers, R_k and S_k, for $1 \le k \le n$ are applied to, respectively, represent the first input and the second input for the OR operation of a bit. Also assume that a one-bit binary number, C_k, for $1 \le k \le n$ is used to represent the output for the OR operation of a bit. For the sake of convenience, suppose that R_k^1 denotes the fact that the value of R_k is 1 and R_k^0 denotes the fact that the value of R_k is 0. Similarly, assume that S_k^1 denotes the fact that the value of S_k is 1 and S_k^0 denotes the fact that the value of S_k is 0. Suppose that C_k^1 denotes the fact that the value of C_k is 1 and C_k^0 denotes the fact that the value of C_k is 0. The following algorithm is proposed to perform the parallel OR operation of a bit.

Algorithm 6.3: ParallelOR(T_0, k)

(1) $T_1 = +(T_0, R_k^1)$ and $T_2 = -(T_0, R_k^1)$.

(2) $T_3 = +(T_1, S_k^1)$ and $T_4 = -(T_1, S_k^1)$.

(3) $T_5 = +(T_2, S_k^1)$ and $T_6 = -(T_2, S_k^1)$.

(4a) **If** (Detect(T_3) = = "yes") **then**

　　　(4) Append-head(T_3, C_k^1).

EndIf

(5a) **If** (Detect(T_4) = = "yes") **then**

　　　(5) Append-head(T_4, C_k^1).

EndIf

(6a) **If** (Detect(T_5) = = "yes") **then**

　　　(6) Append-head(T_5, C_k^1).

EndIf

(7a) **If** (Detect(T_6) = = "yes") **then**

　　　(7) Append-head(T_6, C_k^0).

EndIf

(8) $T_0 = \cup(T_3, T_4, T_5, T_6)$.

EndAlgorithm

Lemma 6-3: *The algorithm,* **ParallelOR**(T_0, k), *can be used to finish the parallel* **OR** *operation of a bit.*

Proof The algorithm, **ParallelOR**(T_0, k), is implemented by means of the *extract, detect, append-head* and *merge* operations. Steps (1) through (3) employ the *extract* operations to form some different test tubes including different inputs (T_1 to T_6). That is, T_1 includes all of the inputs that have $R_k = 1$, T_2 includes all of the inputs that have $R_k = 0$, T_3 includes those that have $R_k = 1$ and $S_k = 1$, T_4 includes those that have $R_k = 1$ and $S_k = 0$, T_5 includes those that have $R_k = 0$ and $S_k = 1$, and finally, T_6 includes those that have $R_k = 0$ and $S_k = 0$. Having performed Steps (1) through (3), this implies that four different inputs for the **OR** operation of a bit as shown in Table 6.2 were poured into tubes T_3 through T_6, respectively.

Steps (4a) and (7a) are, respectively, used to test whether contains any input for tubes T_3, T_4, T_5, and T_6 or not. If any a "yes" is returned for those steps, then the corresponding *append-head* operations will be run. Next, Steps (4) through (7) use the *append-head* operations to append C_k^1 and C_k^0 onto the head of every input in the corresponding tubes. After performing Steps (4) through (7), we can say that four different outputs to the **OR** operation of a bit as shown in Table 6.2 are appended into tubes T_3 through T_6. Finally, the execution of Step (8) applies the *merge* operation to pour tubes T_3 through T_6 into tube T_0. Tube T_0 contains the result performing the **OR** operation of a bit as shown in Table 6.2.　■

6.2.2 The Construction for the Parallel OR Operation of N Bits on Bio-molecular Computer

The parallel **OR** operation of n bits simultaneously generates the corresponding outputs for 2^n combinations of n bits. The following algorithm is proposed to perform the parallel **OR** operation of n bits. Notations in **Algorithm 6.4** are denoted in Sect. 6.2.1.

> **Algorithm 6.4**: **N-Bits-ParallelOR**(T_0)
>
> (1) Append-head(T_1, R_1^1).
>
> (2) Append-head(T_2, R_1^0).
>
> (3) $T_0 = \cup(T_1, T_2)$.
>
> (4) **For** $k = 2$ **to** n
>
>   (4a) Amplify(T_0, T_1, T_2).
>
>   (4b) Append-head(T_1, R_k^1).
>
>   (4c) Append-head(T_2, R_k^0).
>
>   (4d) $T_0 = \cup(T_1, T_2)$.
>
> **EndFor**
>
> (5) **For** $k = 1$ **to** n
>
>   (5a) Append-head(T_0, S_k).
>
> **EndFor**
>
> (6) **For** $k = 1$ **to** n
>
>   (6a) **ParallelOR**(T_0, k).
>
> **EndFor**
>
> **EndAlgorithm**

Lemma 6-4: *The algorithm,* **N-Bits-ParallelOR**(T_0), *can be applied to perform the parallel* **OR** *operation of n bits*

Proof Steps (1) through (4d) are mainly applied to construct solution space of 2^n unsigned integers for the first input (the range of values for them is from 0 to $2^n - 1$). After they are performed, tube T_0 includes those inputs encoding 2^n unsigned integers. Next, Steps (5) through (5a) are used to append the value "0" or "1" for S_k onto the head of solution space of 2^n unsigned integers in tube T_0. Step (6) is the main loop and is mainly used to perform the parallel **OR** operation of n bits. Each execution of Step (6a) calls **ParallelOR**(T_0, k) in Sect. 6.2.1 to perform the **OR** operation for the kth bit of each input in 2^n inputs. Repeat execution of Step (6a) until the nth bit of each input in 2^n inputs is processed. Tube T_0 contains the result performing the parallel **OR** operation of n bits. ∎

6.2.3 The Power for the Parallel OR Operation of N Bits on Bio-molecular Computer

Consider that four values for an unsigned integer of two bits are, subsequently, $00(0_{10})$ $(R_2^0 R_1^0)$, $01(1_{10})$ $(R_2^0 R_1^1)$, $10(2_{10})$ $(R_2^1 R_1^0)$ and $11(3_{10})$ $(R_2^1 R_1^1)$. We want to simultaneously perform the parallel **OR** operation for $11(3_{10})$ $(S_2^1 S_1^1)$ and those four values. **Algorithm 6.4**, **N-Bits-ParallelOR**(T_0), can be used to implement the task. Tube T_0 is an empty tube and is regarded as an input tube of **Algorithm 6.4**. According to Definition 5−2, the input tube T_0 is regarded as the execution environment of the first **BMPDTM**. Similarly, tubes T_1 and T_2 used in **Algorithm 6.4** also are regarded, subsequently, as the execution environment of the second **BMPDTM** and the execution environment of the third **BMPDTM**.

Steps (1) through (4d) in **Algorithm 6.4** are employed to yield a **BMPDTM** with four bio-molecular deterministic one-tape Turing machines. After the execution for Step (1) and Step (2) of **Algorithm 6.4** is implemented, tube $T_1 = \{R_1^1\}$ and tube $T_2 = \{R_1^0\}$. This is to say that a **BMDTM** in the second **BMPDTM** and in the third **BMPDTM** is produced. Figure 6.13 is applied to show the current status of the execution environment to the second **BMPDTM** and the third **BMPDTM**. From Fig. 6.13, the content of the first tape square for the tape in the first **BMDTM** in the second **BMPDTM** is written by its corresponding read-write head and is 1 ($R_1 = 1$), and the content of the first tape square for the tape in the first **BMDTM** in the third **BMPDTM** is written by its corresponding read-write head and is 0 ($R_1 = 0$). For the first **BMDTM** in the second **BMPDTM**, the position of the read-write head is moved to the *left new* tape square, and the state of the finite state control is changed as "$R_1 = 1$". Similarly, for the first **BMDTM** in the third **BMPDTM**, the position of the read-write head is moved to the *left new* tape square, and the state of the finite state control is changed as "$R_1 = 0$".

Next, after the execution for Step (3) of **Algorithm 6.4** is performed, tube $T_0 = \{R_1^1, R_1^0\}$, tube $T_1 = \varnothing$ and tube $T_2 = \varnothing$. This is to say that the execution environment for the first bio-molecular deterministic one-tape Turing machine in the second **BMPDTM** and the first bio-molecular deterministic one-tape Turing machine in the third **BMPDTM** becomes the first **BMPDTM**. The position of the corresponding read-write head and the state of the corresponding finite state control are both reserved. Figure 6.14 is used to illustrate the current status of the execution environment to the first **BMPDTM**. From Fig. 6.14, it is indicated that the contents to the two tapes in the execution environment of the first **BMPDTM** are not changed.

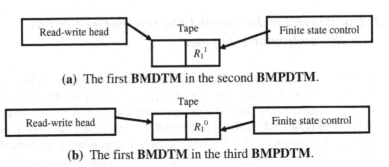

(a) The first **BMDTM** in the second **BMPDTM**.

(b) The first **BMDTM** in the third **BMPDTM**.

Fig. 6.13 Schematic representation of the current status of the execution environment to the second **BMPDTM** and the third **BMPDTM**

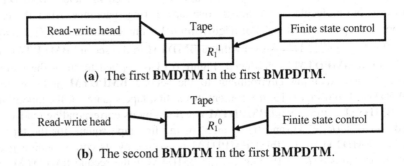

(a) The first **BMDTM** in the first **BMPDTM**.

(b) The second **BMDTM** in the first **BMPDTM**.

Fig. 6.14 Schematic representation of the current status of the execution environment to the first **BMPDTM**

Step (4) is the first loop in **Algorithm 6.4**, the upper bound (n) is two because the number of bits for representing those four values is two. Thus, after the first execution of Step (4a) is finished, tube $T_0 = \varnothing$, tube $T_1 = \{R_1^1, R_1^0\}$ and tube $T_2 = \{R_1^1, R_1^0\}$. This implies that the first **BMDTM** and the second **BMDTM** in the execution environment of the first **BMPDTM** are both copied into the second **BMPDTM** and the third **BMPDTM**. Figure 6.15 is used to show the current status of the execution environment to the second **BMPDTM** and the third **BMPDTM**. From Fig. 6.15, the contents of the first tape square for the corresponding tape of the first **BMDTM** and the corresponding tape of the second **BMDTM** in the execution environment of the second **BMPDTM** are, respectively, 1 ($R_1 = 1$) and 0 ($R_1 = 0$). The contents of the first tape square for the corresponding tape of the first **BMDTM** and the corresponding tape of the second **BMDTM** in the execution environment of the third **BMPDTM** are also, respectively, 0 ($R_1 = 0$) and 1 ($R_1 = 1$). From Fig. 6.15, four bio-molecular deterministic one-tape Turing machines are constructed. For the four bio-molecular deterministic one-tape Turing machines, the position of the corresponding read-write head and the state of the corresponding finite state control are reserved.

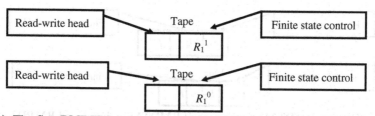

(a) The first **BMDTM** and the second **BMDTM** in the second **BMPDTM**.

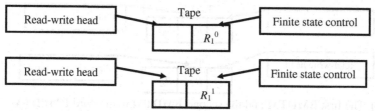

(b) The first **BMDTM** and the second **BMDTM** in the third **BMPDTM**.

Fig. 6.15 Schematic representation of the current status of the execution environment to the second **BMPDTM** and the third **BMPDTM**

Next, after the first execution for Step (4b) and Step (4c) of **Algorithm 6.4** is implemented, tube $T_1 = \{R_2^1 R_1^1, R_2^1 R_1^0\}$ and tube $T_2 = \{R_2^0 R_1^1, R_2^0 R_1^0\}$. This indicates that the content of the *second* tape square for the tape in the first in the second **BMPDTM** is written by its corresponding read-write head and is 1 ($R_2 = 1$), and the content of the *second* tape square for the tape in the second **BMDTM** in the second **BMPDTM** is written by its corresponding read-write head and is also 1 ($R_2 = 1$). Simultaneously, the position of the corresponding read-write head is both moved to the *left new* tape square and the state of the corresponding finite state control is both changed as "$R_2 = 1$". Similarly, the content of the *second* tape square for the tape in the first **BMDTM** in the third **BMPDTM** is written by its corresponding read-write head and is 0 ($R_2 = 0$), and the content of the *second* tape square for the tape in the second **BMDTM** in the third **BMPDTM** is written by its corresponding read-write head and is also 0 ($R_2 = 0$). Simultaneously, the position of the corresponding read-write head is both moved to the *left new* tape square and the state of the corresponding finite state control is both changed as "$R_2 = 0$". Figure 6.16 is applied to explain the current status of the execution environment to the second **BMPDTM** and the third **BMPDTM**.

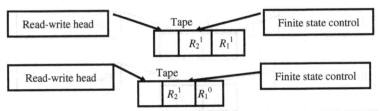

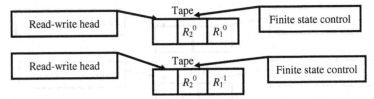

(a) The first **BMDTM** and the second **BMDTM** in the second **BMPDTM**.

(b) The first **BMDTM** and the second **BMDTM** in the third **BMPDTM**.

Fig. 6.16 Schematic representation of the current status of the execution environment to the second **BMPDTM** and the third **BMPDTM**

Next, after the first execution for Step (4d) of **Algorithm 6.4** is performed, tube $T_0 = \{R_2^1 R_1^1, R_2^1 R_1^0, R_2^0 R_1^1, R_2^0 R_1^0\}$, tube $T_1 = \varnothing$ and tube $T_2 = \varnothing$. This implies that the execution environment for those bio-molecular deterministic one-tape Turing machines in the second **BMPDTM** and in the third **BMPDTM** becomes the first **BMPDTM**. Figure 6.17 is employed to illustrate the current status of the execution environment to the first **BMPDTM**. From Fig. 6.17, the contents to the four tapes in the execution environment of the first **BMPDTM** are not changed, and the position of the corresponding read-write head and the state of the corresponding finite state control are reserved.

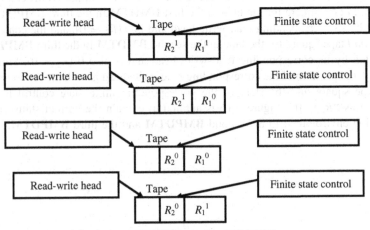

(a) The four **BMDTMs** in the first **BMPDTM**.

Fig. 6.17 Schematic representation of the current status of the execution environment to the first **BMPDTM**

Then, because Step (5) of **Algorithm 6.4** is the second loop and the upper bound for Step (5) of **Algorithm 6.4** is two, Step (5a) will be executed two times. From the first execution and the second execution of Step (5a) in **Algorithm 6.4**, S_1^1 and S_2^1 are, respectively, appended into the head of each bit pattern in tube T_0. This is to say that the contents of the third tape square and the fourth tape square for each tape of the four bio-molecular deterministic one-tape Turing machines in the first **BMPDTM** are written by the corresponding read-write head and are, subsequently, S_1^1 and S_2^1. Simultaneously, for the four bio-molecular deterministic one-tape Turing machines in the first **BMPDTM**, the position of the corresponding read-write head is all moved to the *left new* tape square, the state of the corresponding finite state control is all changed as "$S_2 = 1$" and tube $T_0 = \{S_2^1 S_1^1 R_2^1 R_1^1, S_2^1 S_1^1 R_2^1 R_1^0, S_2^1 S_1^1 R_2^0 R_1^1, S_2^1 S_1^1 R_2^0 R_1^0\}$. Figure 6.18 is used to illustrate the current status of the execution environment to the first **BMPDTM**.

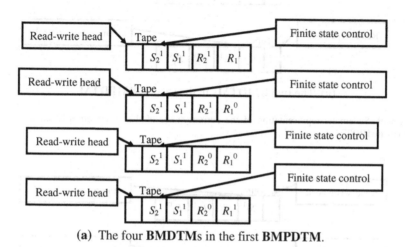

(a) The four **BMDTMs** in the first **BMPDTM**.

Fig. 6.18 Schematic representation of the current status of the execution environment to the first **BMPDTM**

Step (6) in **Algorithm 6.4** is a loop and is applied to perform the parallel **OR** operation of n bits. When the first execution of Step (6a) is implemented, it invokes **Algorithm 6.3** that is used to finish the parallel **OR** operation of one bit, **ParallelOR**(T_0, k), in Sect. 6.2.1. The first parameter, tube T_0, is current the execution environment of the first **BMPDTM** and contains four bio-molecular deterministic one-tape Turing machines (Fig. 6.18). It is regarded as an input tube of **Algorithm 6.3**. The value for the second parameter, k, is one and is also regarded as an input value of **Algorithm 6.3**.

When **Algorithm 6.3** is first called, seven tubes are all regarded as independent environments of seven bio-molecular parallel deterministic one-tape Turing machines. Tube T_0 is regarded as the first **BMPDTM** and tube T_k is regarded as the

$(k + 1)$th **BMPDTM**. After the *first* execution of Step (1) in **Algorithm 6.3** is performed, tube $T_0 = \emptyset$, tube $T_1 = \{S_2^1 S_1^1 R_2^1 R_1^1, S_2^1 S_1^1 R_2^0 R_1^1\}$ and tube $T_2 = \{S_2^1 S_1^1 R_2^1 R_1^0, S_2^1 S_1^1 R_2^0 R_1^0\}$. This implies that the new execution environments for two bio-molecular deterministic one-tape Turing machines with the content of tape square, "R_1^1", and other two bio-molecular deterministic one-tape Turing machines with the content of tape square, "R_1^0" are, respectively, the second **BMPDTM** and the third **BMPDTM**. The position of the corresponding read-write head and the state of the corresponding finite state control are reserved. Figures 6.19 and 6.20 are used to explain the result.

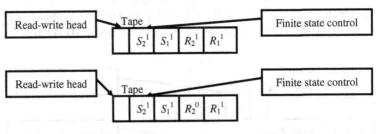

(a) The two **BMDTMs** in the second **BMPDTM**.

Fig. 6.19 Schematic representation of the current status of the execution environment to the second **BMPDTM**

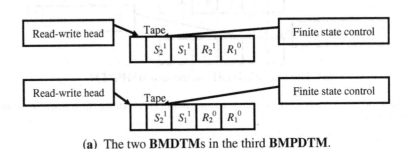

(a) The two **BMDTMs** in the third **BMPDTM**.

Fig. 6.20 Schematic representation of the current status of the execution environment to the third **BMPDTM**

Then, after the *first* execution of Steps (2) and (3) in **Algorithm 6.3** is implemented, tube $T_1 = \emptyset$, tube $T_2 = \emptyset$, tube $T_4 = \emptyset$, tube $T_6 = \emptyset$, tube $T_3 = \{S_2^1 S_1^1 R_2^1 R_1^1, S_2^1 S_1^1 R_2^0 R_1^1\}$ and tube $T_5 = \{S_2^1 S_1^1 R_2^1 R_1^0, S_2^1 S_1^1 R_2^0 R_1^0\}$. This is to say that the new execution environments for two bio-molecular deterministic one-tape Turing machines with the contents of tape square, "R_1^1" and "S_1^1", and other two bio-molecular deterministic one-tape Turing machines with the content of tape square, "R_1^0" and "S_1^1" are, respectively, the fourth **BMPDTM** and the sixth **BMPDTM**. The position of the corresponding read-write head and the state of the corresponding finite state control are reserved. Figures 6.21 and 6.22 are applied to illustrate the result.

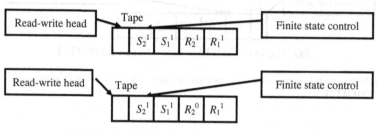

(a) The two **BMDTMs** in the fourth **BMPDTM**.

Fig. 6.21 Schematic representation of the current status of the execution environment to the fourth **BMPDTM**

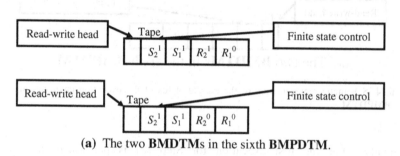

(a) The two **BMDTMs** in the sixth **BMPDTM**.

Fig. 6.22 Schematic representation of the current status of the execution environment to the sixth **BMPDTM**

Next, after the first execution for Steps (4a), (5a), (6a) and (7a) is performed, the returned result from Steps (4a) and (6a) is "yes". Therefore, after the first execution for Steps (4) and (6) is implemented, tube $T_3 = \{C_1^1 S_2^1 S_1^1 R_2^1 R_1^1, C_1^1 S_2^1 S_1^1 R_2^0 R_1^1\}$ and tube $T_5 = \{C_1^1 S_2^1 S_1^1 R_2^1 R_1^0, C_1^1 S_2^1 S_1^1 R_2^0 R_1^0\}$. This indicates that the content of the fifth tape square for each tape in the two **BMDTMs** of the fourth **BMPDTM** and the two **BMDTMs** of the sixth **BMPDTM** is all written by the corresponding read-write head and is 1 ($C_1 = 1$). Simultaneously, for the two **BMDTMs** of the fourth **BMPDTM** and the two **BMDTMs** of the sixth **BMPDTM**, the state of each finite state control is changed as "$C_1 = 1$" and the position of each read-write head is moved to the *left new* tape square of the corresponding tape. Figures 6.23 and 6.24 are used to respectively illustrate the current status of the execution environment to the fourth **BMPDTM** and the sixth **BMPDTM**.

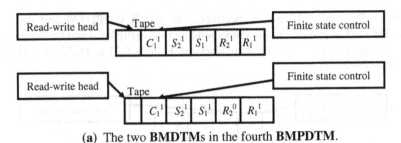

(a) The two **BMDTMs** in the fourth **BMPDTM**.

Fig. 6.23 Schematic representation of the current status of the execution environment to the fourth **BMPDTM**

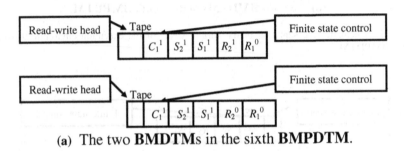

(a) The two **BMDTMs** in the sixth **BMPDTM**.

Fig. 6.24 Schematic representation of the current status of the execution environment to the sixth **BMPDTM**

Finally, after the first execution for Step (8) is performed, tube $T_0 = \{C_1^1 S_2^1 S_1^1 R_2^1 R_1^1, C_1^1 S_2^1 S_1^1 R_2^0 R_1^1, C_1^1 S_2^1 S_1^1 R_2^1 R_1^0, C_1^1 S_2^1 S_1^1 R_2^0 R_1^0\}$, tube $T_1 = \varnothing$, tube $T_2 = \varnothing$, $T_3 = \varnothing$, $T_4 = \varnothing$, $T_5 = \varnothing$ and tube $T_6 = \varnothing$. This is to say that the execution environment for the first and second bio-molecular deterministic one-tape Turing machines in the fourth **BMPDTM** and the first and second bio-molecular deterministic one-tape Turing machines in the sixth **BMPDTM** becomes the first **BMPDTM**. The position of the corresponding read-write head and the state of the corresponding finite state control are all reserved. Figure 6.25 is used to explain the result.

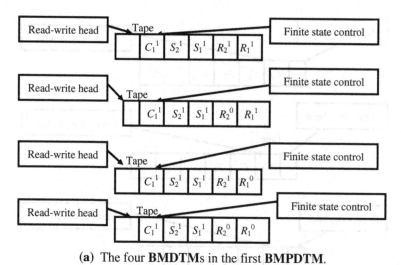

(a) The four **BMDTMs** in the first **BMPDTM**.

Fig. 6.25 Schematic representation of the current status of the execution environment to the first **BMPDTM**

After the execution of the *first* time for each operation in **Algorithm 6.3** is implemented, the parallel **OR** operation to the first bit of the two inputs is also performed. Next, when the *second* execution of Step (6a) in **Algorithm 6.4** is finished, it again invokes **Algorithm 6.3**. The first parameter, tube T_0, is current the execution environment of the first **BMPDTM** and contains four bio-molecular deterministic one-tape Turing machines (Fig. 6.25). It is regarded as an input tube of **Algorithm 6.3**. The value for the second parameter, k, is *two* and is also regarded as an input value of **Algorithm 6.3**. After the execution of the *second* time for each operation in **Algorithm 6.3** is performed, tube $T_0 = \{ C_2^1 C_1^1 S_2^1 S_1^1 R_2^1 R_1^1, \ C_2^1 C_1^1 S_2^1 S_1^1 R_2^0 R_1^1, \ C_2^1 C_1^1 S_2^1 S_1^1 R_2^1 R_1^0, \ C_2^1 C_1^1 S_2^1 S_1^1 R_2^0 R_1^0 \}$ and other tubes become all empty tubes. This implies that the content of the sixth tape square for each **BMDTM** is written by the corresponding read-write head and is C_2^1. Simultaneously, the state of each finite state control is changed as "$C_2 = 1$" and the position of each read-write head is moved to the *left new* tape square of the corresponding tape. Figure 6.26 is used to show the result and **Algorithm 6.4** is terminated.

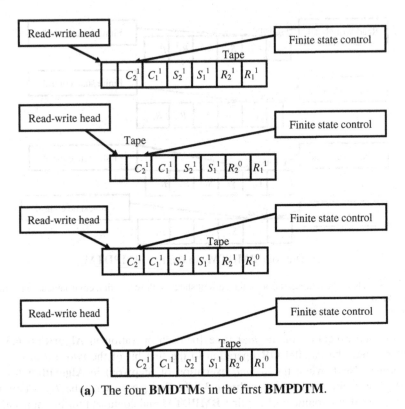

(a) The four **BMDTMs** in the first **BMPDTM**.

Fig. 6.26 Schematic representation of the current status of the execution environment to the first **BMPDTM**

6.3 Introduction to AND Operation on Bio-molecular Computer

The **AND** operation of a bit for two Boolean variables R and S generates a result of 1 if both R and S are 1. However, if either R or S, or both, are zero, then the result is 0. The dot $\cdot$ and $\wedge$ symbol are both used to represent the **AND** operation. The four possible combinations for the **AND** operation of a bit to two Boolean variables R and S are

$$0 \wedge 0 = 0$$
$$0 \wedge 1 = 0$$
$$1 \wedge 0 = 0$$
$$1 \wedge 1 = 1$$

A truth table is usually applied with logic operation to represent all possible combinations of inputs and the corresponding outputs. The truth table for the **AND** operation is shown in Table 6.3.

Table 6.3 The truth table for the **AND** operation of a bit for two Boolean variables R and S

Input		Output
R	S	$C = R \wedge S$
0	0	0
0	1	0
1	0	0
1	1	1

6.3.1 The Construction for the Parallel AND Operation of a Bit on Bio-molecular Computer

Assume that two one-bit binary numbers, R_k and S_k, for $1 \leq k \leq n$ are employed to, respectively, represent the first input and the second input for the **AND** operation of a bit. Also assume that a one-bit binary number, C_k, for $1 \leq k \leq n$ is applied to represent the output for the **AND** operation of a bit. For the sake of convenience, suppose that R_k^1 denotes the fact that the value of R_k is 1 and R_k^0 denotes the fact that the value of R_k is 0. Similarly, assume that S_k^1 denotes the fact that the value of S_k is 1 and S_k^0 denotes the fact that the value of S_k is 0. Suppose that C_k^1 denotes the fact that the value of C_k is 1 and C_k^0 denotes the fact that the value of C_k is 0. The following algorithm is offered to perform the parallel **AND** operation of a bit.

(1) $T_1 = +(T_0, R_k^1)$ and $T_2 = -(T_0, R_k^1$

(2) $T_3 = +(T_1, S_k^1)$ and $T_4 = -(T_1, S_k^1$

(3) $T_5 = +(T_2, S_k^1)$ and $T_6 = -(T_2, S_k^1$

(4a) **If** $(\text{Detect}(T_3) == \text{"yes"})$ **then**

    (4) Append-head(T_3, C_k^1).

EndIf

(5a) **If** $(\text{Detect}(T_4) == \text{"yes"})$ **then**

    (5) Append-head(T_4, C_k^0).

EndIf

(6a) **If** $(\text{Detect}(T_5) == \text{"yes"})$ **then**

    (6) Append-head(T_5, C_k^0).

EndIf

(7a) **If** $(\text{Detect}(T_6) == \text{"yes"})$ **then**

    (7) Append-head(T_6, C_k^0).

EndIf

(8) $T_0 = \cup(T_3, T_4, T_5, T_6)$.

EndAlgorithm

Lemma 6-5: *The algorithm,* **ParallelAND**(T_0, k), *can be used to perform the parallel* **AND** *operation of a bit.*

Proof The algorithm, **ParallelAND**(T_0, k), is implemented by means of the *extract, detect, append-head* and *merge* operations. Steps (1) through (3) employ the *extract* operations to form some different test tubes including different inputs (T_1 to T_6). That is, T_1 includes all of the inputs that have $R_k = 1$, T_2 includes all of the inputs that have $R_k = 0$, T_3 includes those that have $R_k = 1$ and $S_k = 1$, T_4 includes those that have $R_k = 1$ and $S_k = 0$, T_5 includes those that have $R_k = 0$ and $S_k = 1$, and finally, T_6 includes those that have $R_k = 0$ and $S_k = 0$. Having finished Steps (1) through (3), this implies that four different inputs for the **AND** operation of a bit as shown in Table 6.3 were poured into tubes T_3 through T_6, respectively.

Steps (4a) and (7a) are, respectively, used to test whether contains any input for tubes T_3, T_4, T_5, and T_6 or not. If any a "yes" is returned for those steps, then the corresponding *append-head* operations will be run. Next, Steps (4) through (7) use the *append-head* operations to append C_k^1 and C_k^0 onto the head of every input in the corresponding tubes. After performing Steps (4) through (7), we can say that four different outputs to the **AND** operation of a bit as shown in Table 6.3 are appended into tubes T_3 through T_6. Finally, the execution of Step (8) applies the *merge* operation to pour tubes T_3 through T_6 into tube T_0. Tube T_0 contains the result performing the **AND** operation of a bit as shown in Table 6.3.   ∎

6.3.2 The Construction for the Parallel AND Operation of N Bits on Bio-molecular Computer

The parallel **AND** operation of n bits simultaneously generates the corresponding outputs for 2^n combinations of n bits. The following algorithm is offered to finish the parallel **AND** operation of n bits. Notations in **Algorithm 6.6** are denoted in Sect. 6.3.1.

Algorithm 6.6: **N-Bits-ParallelAND**(T_0)

(1) Append-head(T_1, R_1^1).

(2) Append-head(T_2, R_1^0).

(3) $T_0 = \cup(T_1, T_2)$.

(4) **For** $k = 2$ **to** n

    (4a) Amplify(T_0, T_1, T_2).

    (4b) Append-head(T_1, R_k^1).

    (4c) Append-head(T_2, R_k^0).

    (4d) $T_0 = \cup(T_1, T_2)$.

EndFor

(5) **For** $k = 1$ **to** n

    (5a) Append-head(T_0, S_k).

EndFor

(6) **For** $k = 1$ **to** n

    (6a) **ParallelAND**(T_0, k).

EndFor

EndAlgorithm

Lemma 6-6: *The algorithm,* **N-Bits-ParallelAND**(T_0), *can be employed to perform the parallel* **AND** *operation of n bits.*

Proof Steps (1) through (4d) are mainly used to construct solution space of 2^n unsigned integers for the first input (the range of values for them is from 0 to 2^n-1). After they are performed, tube T_0 includes those inputs encoding 2^n unsigned integers. Next, Steps (5) through (5a) are used to append the value "0" or "1" for S_k onto the head of solution space of 2^n unsigned integers in tube T_0. Step (6) is the main loop and is mainly used to perform the parallel **AND** operation of n bits. Each execution of Step (6a) calls **ParallelAND**(T_0, k) in Sect. 6.3.1 to perform the **AND** operation for the kth bit of each input in 2^n inputs. Repeat execution of Step (6a) until the nth bit of each input in 2^n inputs is processed. Tube T_0 contains the result performing the parallel **AND** operation of n bits.   ■

6.3.3 The Power for the Parallel AND Operation of N Bits on Bio-molecular Computer

Consider that four values for an unsigned integer of two bits are, subsequently, $00(0_{10})$ $\left(R_2^0 R_1^0\right)$, $01(1_{10})$ $\left(R_2^0 R_1^1\right)$, $10(2_{10})$ $\left(R_2^1 R_1^0\right)$ and $11(3_{10})$ $\left(R_2^1 R_1^1\right)$. We want to simultaneously perform the parallel **AND** operation for $00(0_{10})$ $\left(S_2^0 S_1^0\right)$ and those four values. **Algorithm 6.6**, **N-Bits-ParallelAND**(T_0), can be applied to perform the task. Tube T_0 is an empty tube and is regarded as an input tube of **Algorithm 6.6**. Due to Definition 5–2, the input tube T_0 is regarded as the execution environment of the first **BMPDTM**. Similarly, tubes T_1 and T_2 used in **Algorithm 6.6** also are regarded, subsequently, as the execution environment of the second **BMPDTM** and the execution environment of the third **BMPDTM**.

Steps (1) through (4d) in **Algorithm 6.6** are used to construct a **BMPDTM** with four bio-molecular deterministic one-tape Turing machines. After the execution for Step (1) and Step (2) of **Algorithm 6.6** is finished, tube $T_1 = \left\{R_1^1\right\}$ and tube $T_2 = \left\{R_1^0\right\}$. This implies that a **BMDTM** in the second **BMPDTM** and in the third **BMPDTM** is generated. Figure 6.27 is employed to illustrate the current status of the execution environment to the second **BMPDTM** and the third **BMPDTM**. From Fig. 6.27, the content of the first tape square for the tape in the first **BMDTM** in the second **BMPDTM** is written by its corresponding read-write head and is 1 ($R_1 = 1$), and the content of the first tape square for the tape in the first **BMDTM** in the third **BMPDTM** is written by its corresponding read-write head and is 0 ($R_1 = 0$). For the first **BMDTM** in the second **BMPDTM**, the position of the read-write head is moved to the *left new* tape square, and the state of the finite state control is changed as "$R_1 = 1$". Similarly, for the first **BMDTM** in the third **BMPDTM**, the position of the read-write head is also moved to the *left new* tape square, and the state of the finite state control is changed as "$R_1 = 0$".

Next, after the execution for Step (3) of **Algorithm 6.6** is implemented, tube $T_0 = \{R_1^1, R_1^0\}$, tube $T_1 = \varnothing$ and tube $T_2 = \varnothing$. This indicates that the execution environment for the first bio-molecular deterministic one-tape Turing machine in the second **BMPDTM** and the first bio-molecular deterministic one-tape Turing machine in the third **BMPDTM** becomes the first **BMPDTM**. The position of the corresponding read-write head and the state of the corresponding finite state control are both reserved. Figure 6.28 is applied to show the current status of the execution environment to the first **BMPDTM**. From Fig. 6.28, it is indicated that the contents to the two tapes in the execution environment of the first **BMPDTM** are not changed.

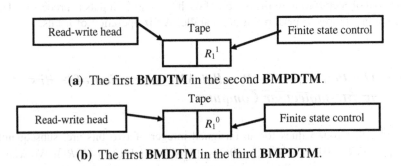

(a) The first **BMDTM** in the second **BMPDTM**.

(b) The first **BMDTM** in the third **BMPDTM**.

Fig. 6.27 Schematic representation of the current status of the execution environment to the second **BMPDTM** and the third **BMPDTM**

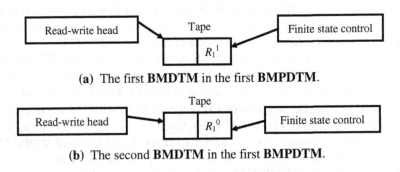

(a) The first **BMDTM** in the first **BMPDTM**.

(b) The second **BMDTM** in the first **BMPDTM**.

Fig. 6.28 Schematic representation of the current status of the execution environment to the first **BMPDTM**

Step (4) is the first loop in **Algorithm 6.6**, the upper bound (n) is two because the number of bits for representing those four values is two. Therefore, after the first execution of Step (4a) is performed, tube $T_0 = \varnothing$, tube $T_1 = \{R_1^1, R_1^0\}$ and tube $T_2 = \{R_1^1, R_1^0\}$. This is to say that the first **BMDTM** and the second **BMDTM** in the execution environment of the first **BMPDTM** are both copied into the second **BMPDTM** and the third **BMPDTM**. Figure 6.29 is employed to explain the current status of the execution environment to the second **BMPDTM** and the third **BMPDTM**. From Fig. 6.29, the contents of the first tape square for the corresponding tape of the first **BMDTM** and the corresponding tape of the second **BMDTM** in the execution environment of the second **BMPDTM** are, respectively, 1 ($R_1 = 1$) and 0 ($R_1 = 0$). The contents of the first tape square for the corresponding tape of the first **BMDTM** and the corresponding tape of the second **BMDTM** in the execution environment of the third **BMPDTM** are also, respectively, 0 ($R_1 = 0$) and 1 ($R_1 = 1$). From Fig. 6.29, four **BMDTMs** are produced. For the four **BMDTMs**, the position of the corresponding read-write head and the state of the corresponding finite state control are reserved.

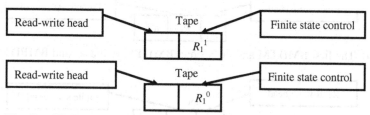

(a) The first **BMDTM** and the second **BMDTM** in the second **BMPDTM**.

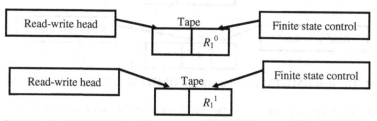

(b) The first **BMDTM** and the second **BMDTM** in the third **BMPDTM**.

Fig. 6.29 Schematic representation of the current status of the execution environment to the second **BMPDTM** and the third **BMPDTM**

Next, after the first execution for Step (4b) and Step (4c) in **Algorithm 6.6** is performed, tube $T_1 = \{R_2^1 R_1^1, R_2^1 R_1^0\}$ and tube $T_2 = \{R_2^0 R_1^1, R_2^0 R_1^0\}$. This implies that the content of the *second* tape square for the tape in the first **BMDTM** in the second **BMPDTM** is written by its corresponding read-write head and is 1 ($R_2 = 1$), and the content of the *second* tape square for the tape in the second **BMDTM** in the second **BMPDTM** is written by its corresponding read-write head and is also 1 ($R_2 = 1$). Simultaneously, the position of the corresponding read-write head is both moved to the *left new* tape square and the state of the corresponding finite state control is both changed as "$R_2 = 1$". Similarly, the content of the *second* tape square for the tape in the first **BMDTM** in the third **BMPDTM** is written by its corresponding read-write head and is 0 ($R_2 = 0$), and the content of the *second* tape square for the tape in the second **BMDTM** in the third **BMPDTM** is written by its corresponding read-write head and is also 0 ($R_2 = 0$). Simultaneously, the position of the corresponding read-write head is both moved to the *left new* tape square and the state of the corresponding finite state control is both changed as "$R_2 = 0$". Figure 6.30 is used to show the current status of the execution environment to the second **BMPDTM** and the third **BMPDTM**.

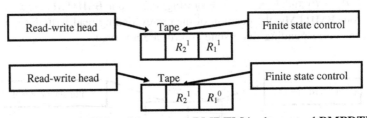

(a) The first **BMDTM** and the second **BMDTM** in the second **BMPDTM**.

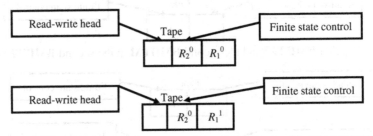

(b) The first **BMDTM** and the second **BMDTM** in the third **BMPDTM**.

Fig. 6.30 Schematic representation of the current status of the execution environment to the second **BMPDTM** and the third **BMPDTM**

Next, after the first execution for Step (4d) of **Algorithm 6.6** is implemented, tube $T_0 = \{R_2^1 R_1^1, R_2^1 R_1^0, R_2^0 R_1^1, R_2^0 R_1^0\}$, tube $T_1 = \varnothing$ and tube $T_2 = \varnothing$. This is to say that the execution environment for those bio-molecular deterministic one-tape Turing machines in the second **BMPDTM** and in the third **BMPDTM** becomes the first **BMPDTM**. Figure 6.31 is used to explain the current status of the execution environment to the first **BMPDTM**. From Fig. 6.31, the contents to the four tapes in the execution environment of the first **BMPDTM** are not changed, and the position of the corresponding read-write head and the state of the corresponding finite state control are reserved.

Then, because Step (5) of **Algorithm 6.6** is the second loop and the upper bound for Step (5) of **Algorithm 6.6** is two, Step (5a) will be executed two times. From the first execution and the second execution of Step (5a) in **Algorithm 6.6**, S_1^0 and S_2^0 are, respectively, appended into the head of each bit pattern in tube T_0.

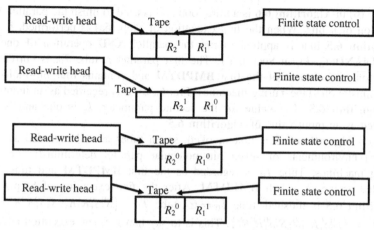

(a) The four **BMDTM**s in the first **BMPDTM**.

Fig. 6.31 Schematic representation of the current status of the execution environment to the first **BMPDTM**

This indicates that the contents of the third tape square and the fourth tape square for each tape of the four bio-molecular deterministic one-tape Turing machines in the first **BMPDTM** are written by the corresponding read-write head and are, subsequently, S_1^0 and S_2^0. Simultaneously, for the four bio-molecular deterministic one-tape Turing machines in the first **BMPDTM**, the position of the corresponding read-write head is all moved to the *left new* tape square, the state of the corresponding finite state control is all changed as "$S_2 = 0$" and tube $T_0 = \{S_2^0 S_1^0 R_2^1 R_1^1, S_2^0 S_1^0 R_2^1 R_1^0, S_2^0 S_1^0 R_2^0 R_1^1, S_2^0 S_1^0 R_2^0 R_1^0\}$. Figure 6.32 is applied to show the current status of the execution environment to the first **BMPDTM**.

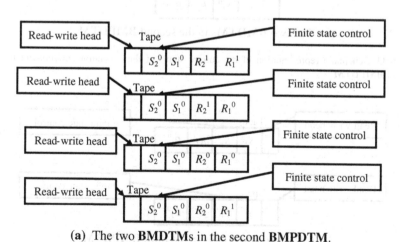

(a) The two **BMDTM**s in the second **BMPDTM**.

Fig. 6.32 Schematic representation of the current status of the execution environment to the first **BMPDTM**

Step (6) in **Algorithm 6.6** is a third loop and is used to finish the parallel **AND** operation of n bits. When the first execution of Step (6a) is performed, it calls **Algorithm 6.5** that is applied to run the parallel **AND** operation of one bit, **ParallelAND**(T_0, k), in Sect. 6.3.1. The first parameter, tube T_0, is current the execution environment of the first **BMPDTM** and contains four bio-molecular deterministic one-tape Turing machines (Fig. 6.32). It is regarded as an input tube of **Algorithm 6.5**. The value for the second parameter, k, is one and is also regarded as an input value of **Algorithm 6.5**.

When **Algorithm 6.5** is first invoked, seven tubes are all regarded as independent environments of seven bio-molecular parallel deterministic one-tape Turing machines. Tube T_0 is regarded as the first **BMPDTM** and tube T_k is regarded as the $(k + 1)$th **BMPDTM**. After the *first* execution of Step (1) in **Algorithm 6.5** is finished, tube $T_0 = \varnothing$, tube $T_1 = \{S_2^0 S_1^0 R_2^1 R_1^1, S_2^0 S_1^0 R_2^0 R_1^1\}$ and tube $T_2 = \{S_2^0 S_1^0 R_2^1 R_1^0, S_2^0 S_1^0 R_2^0 R_1^0\}$. This is to say that the new execution environments for two bio-molecular deterministic one-tape Turing machines with the content of tape square, "R_1^1", and other two bio-molecular deterministic one-tape Turing machines with the content of tape square, "R_1^0" are, respectively, the second **BMPDTM** and the third **BMPDTM**. The position of the corresponding read-write head and the state of the corresponding finite state control are reserved. Figures 6.33 and 6.34 are applied to illustrate the result.

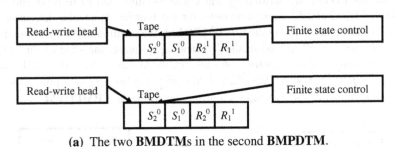

(a) The two **BMDTMs** in the second **BMPDTM**.

Fig. 6.33 Schematic representation of the current status of the execution environment to the second **BMPDTM**

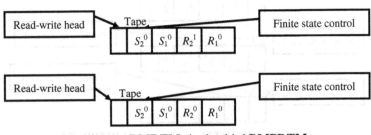

(a) The two **BMDTMs** in the third **BMPDTM**.

Fig. 6.34 Schematic representation of the current status of the execution environment to the third **BMPDTM**

Then, after the *first* execution of Steps (2) and (3) in **Algorithm 6.5** is performed, tube $T_1 = \varnothing$, tube $T_2 = \varnothing$, tube $T_3 = \varnothing$, tube $T_5 = \varnothing$, tube $T_4 = \{S_2^0 S_1^0 R_2^1 R_1^1, S_2^0 S_1^0 R_2^0 R_1^1\}$ and tube $T_6 = \{S_2^0 S_1^0 R_2^1 R_1^0, S_2^0 S_1^0 R_2^0 R_1^0\}$. This implies that the new execution environments for two bio-molecular deterministic one-tape Turing machines with the contents of tape square, "R_1^1" and "S_1^0", and other two bio-molecular deterministic one-tape Turing machines with the content of tape square, "R_1^0" and "S_1^0" are, respectively, the fifth **BMPDTM** and the seventh **BMPDTM**. The position of the corresponding read-write head and the state of the corresponding finite state control are reserved. Figures 6.35 and 6.36 are used to show the result.

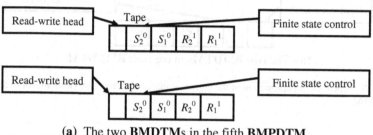

(a) The two **BMDTMs** in the fifth **BMPDTM**.

Fig. 6.35 Schematic representation of the current status of the execution environment to the fifth **BMPDTM**

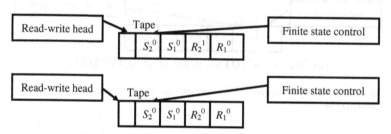

(a) The two **BMDTMs** in the seventh **BMPDTM**.

Fig. 6.36 Schematic representation of the current status of the execution environment to the seventh **BMPDTM**

Next, after the first execution for Steps (4a), (5a), (6a) and (7a) is finished, the returned result from Steps (5a) and (7a) is "yes". Therefore, after the first execution for Steps (5) and (7) is performed, tube $T_4 = \{C_1^0 S_2^0 S_1^0 R_2^1 R_1^1, C_1^0 S_2^0 S_1^0 R_2^0 R_1^1\}$ and tube $T_6 = \{C_1^0 S_2^0 S_1^0 R_2^1 R_1^0, C_1^0 S_2^0 S_1^0 R_2^0 R_1^0\}$. This is to say that the content of the fifth tape square for each tape in the two **BMDTMs** of the fifth **BMPDTM** and the two **BMDTMs** of the seventh **BMPDTM** is all written by the corresponding read-write head and is 0 ($C_1 = 0$). Simultaneously, for the two **BMDTMs** of the fifth **BMPDTM** and the two **BMDTMs** of the seventh **BMPDTM**, the state of each finite state control is changed as "$C_1 = 0$" and the position of each read-write head is moved to the *left new* tape square of the corresponding tape. Figures 6.37 and 6.38 are applied to subsequently explain the current status of the execution environment to the fifth **BMPDTM** and the seventh **BMPDTM**.

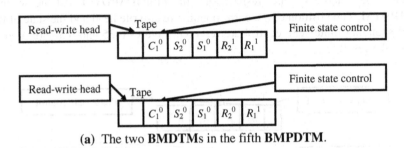

(a) The two **BMDTMs** in the fifth **BMPDTM**.

Fig. 6.37 Schematic representation of the current status of the execution environment to the fifth **BMPDTM**

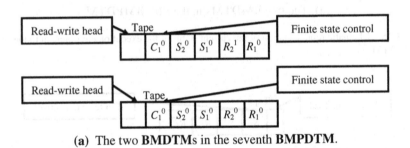

(a) The two **BMDTMs** in the seventh **BMPDTM**.

Fig. 6.38 Schematic representation of the current status of the execution environment to the seventh **BMPDTM**

Finally, after the first execution for Step (8) is finished, tube $T_0 = \{C_1^0 S_2^0 S_1^0 R_2^1 R_1^1, C_1^0 S_2^0 S_1^0 R_2^0 R_1^1, C_1^0 S_2^0 S_1^0 R_2^1 R_1^0, C_1^0 S_2^0 S_1^0 R_2^0 R_1^0\}$, tube $T_1 = \varnothing$, tube $T_2 = \varnothing$, $T_3 = \varnothing$, $T_4 = \varnothing$, $T_5 = \varnothing$ and tube $T_6 = \varnothing$. This indicates that the execution environment for the first and second bio-molecular deterministic one-tape Turing machines in the fifth **BMPDTM** and the first and second bio-molecular deterministic one-tape Turing machines in the seventh **BMPDTM** becomes the first **BMPDTM**. The position of the corresponding read-write head and the state of the corresponding finite state control are all reserved. Figure 6.39 is used to explain the result.

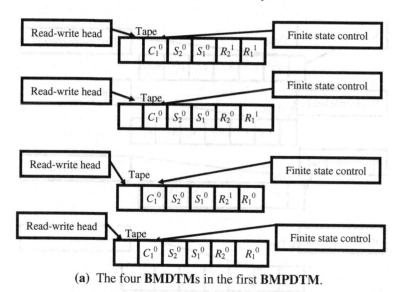

(a) The four **BMDTMs** in the first **BMPDTM**.

Fig. 6.39 Schematic representation of the current status of the execution environment to the first **BMPDTM**

After the execution of the *first* time for each operation in **Algorithm 6.5** is performed, the parallel **AND** operation to the first bit of the two inputs is also finished. Next, when the *second* execution of Step (6a) in **Algorithm 6.6** is implemented, it again invokes **Algorithm 6.5**. The first parameter, tube T_0, is current the execution environment of the first **BMPDTM** and contains four bio-molecular deterministic one-tape Turing machines (Fig. 6.39). It is regarded as an input tube of **Algorithm 6.5**. The value for the second parameter, k, is *two* and is also regarded as an input value of **Algorithm 6.5**. After the execution of the *second* time for each operation in **Algorithm 6.5** is implemented, tube $T_0 = \{C_2^0 C_1^0 S_2^0 S_1^0 R_2^1 R_1^1, C_2^0 C_1^0 S_2^0 S_1^0 R_2^0 R_1^1, C_2^0 C_1^0 S_2^0 S_1^0 R_2^1 R_1^0, C_2^0 C_1^0 S_2^0 S_1^0 R_2^0 R_1^0\}$ and other tubes become all empty tubes. This is to say that the content of the sixth tape square for each **BMDTM** is written by the corresponding read-write head and is 1 ($C_2 = 1$). Simultaneously, the state of each finite state control is changed as "$C_2 = 1$" and the position of each read-write head is moved to the *left new* tape square of the corresponding tape. Figure 6.40 is employed to illustrate the result and **Algorithm 6.6** is terminated.

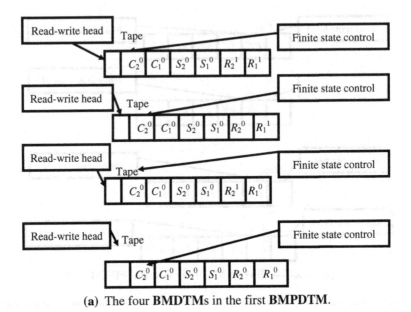

(a) The four **BMDTMs** in the first **BMPDTM**.

Fig. 6.40 Schematic representation of the current status of the execution environment to the first **BMPDTM**

6.4 Introduction for NOR Operation on Bio-molecular Computer

The **NOR** operation of a bit for two Boolean variables R and S generates a result of 1 if both R and S are 0. However, if either R or S, or both, are 1, then the result is 0. The four possible combinations for the **NOR** operation of a bit to two Boolean variables R and S are

$$\overline{0 \vee 0} = 1$$
$$\overline{0 \vee 1} = 0$$
$$\overline{1 \vee 0} = 0$$
$$\overline{1 \vee 1} = 0$$

A truth table is usually applied with logic operation to represent all possible combinations of inputs and the corresponding outputs. The truth table for the **NOR** operation is shown in Table 6.4.

Table 6.4 The truth table for the **NOR** operation of a bit for two Boolean variables R and S

Input		Output
R	S	$C = \overline{R \vee S}$
0	0	1
0	1	0
1	0	0
1	1	0

6.4.1 The Construction for the Parallel NOR Operation of a Bit on Bio-molecular Computer

Suppose that two one-bit binary numbers, R_k and S_k, for $1 \leq k \leq n$ are employed to, respectively, represent the first input and the second input for the **NOR** operation of a bit. Also assume that a one-bit binary number, C_k, for $1 \leq k \leq n$ is applied to represent the output for the **NOR** operation of a bit. For the sake of convenience, suppose that R_k^1 denotes the fact that the value of R_k is 1 and R_k^0 denotes the fact that the value of R_k is 0. Similarly, assume that S_k^1 denotes the fact that the value of S_k is 1 and S_k^0 denotes the fact that the value of S_k is 0. Suppose that C_k^1 denotes the fact that the value of C_k is 1 and C_k^0 denotes the fact that the value of C_k is 0. The following algorithm is offered to perform the parallel **NOR** operation of a bit.

Algorithm 6.7: ParallelNOR(T_0, k)

(1) $T_1 = +(T_0, R_k^1)$ and $T_2 = -(T_0, R_k^1)$.

(2) $T_3 = +(T_1, S_k^1)$ and $T_4 = -(T_1, S_k^1)$.

(3) $T_5 = +(T_2, S_k^1)$ and $T_6 = -(T_2, S_k^1)$.

(4a) **If** (Detect(T_3) $==$ "yes") **then**

    (4) Append-head(T_3, C_k^0).

EndIf

(5a) **If** (Detect(T_4) $==$ "yes") **then**

    (5) Append-head(T_4, C_k^0).

EndIf

(6a) **If** (Detect(T_5) $==$ "yes") **then**

    (6) Append-head(T_5, C_k^0).

EndIf

(7a) **If** (Detect(T_6) $==$ "yes") **then**

    (7) Append-head(T_6, C_k^1).

EndIf

(8) $T_0 = \cup(T_3, T_4, T_5, T_6)$.

EndAlgorithm

Lemma 6-7: *The algorithm,* **ParallelNOR(T_0, k),** *can be used to finish the parallel* **NOR** *operation of a bit.*

Proof The algorithm, **ParallelNOR(T_0, k)**, is implemented by means of the *extract*, *detect*, *append-head* and *merge* operations. Steps (1) through (3) employ the *extract* operations to form some different tubes containing different inputs (T_1 to T_6). That is, T_1 consists of all of the inputs that have $R_k = 1$, T_2 includes all of the inputs that have $R_k = 0$, T_3 contains those that have $R_k = 1$ and $S_k = 1$, T_4 consists of those that have $R_k = 1$ and $S_k = 0$, T_5 includes those that have $R_k = 0$ and $S_k = 1$, and finally, T_6 contains those that have $R_k = 0$ and $S_k = 0$. Having performed Steps (1) through (3), this is to say that four different inputs for the

NOR operation of a bit as shown in Table 6.4 were poured into tubes T_3 through T_6, respectively.

Steps (4a) and (7a) are, respectively, used to test whether contains any input for tubes T_3, T_4, T_5, and T_6 or not. If any a "yes" is returned for those steps, then the corresponding *append-head* operations will be run. Because tubes T_3, T_4, T_5 and T_6, subsequently, contains the input for the fourth row, the third row, the second row and the first row in Table 6.4, Steps (4) through (7) apply the *append-head* operations to append C_k^0 or C_k^1 onto the head of every input in the corresponding tubes. After performing Steps (4) through (7), we can say that four different outputs to the **NOR** operation of a bit as shown in Table 6.4 are appended into tubes T_3 through T_6. Finally, the execution of Step (8) uses the *merge* operation to pour tubes T_3 through T_6 into tube T_0. Tube T_0 contains the result finishing the **NOR** operation of a bit as shown in Table6.4.   ∎

6.4.2 The Construction for the Parallel NOR Operation of N Bits on Bio-molecular Computer

The parallel **NOR** operation of n bits simultaneously yields the corresponding outputs for 2^n combinations of n bits. The following algorithm is proposed to perform the parallel **NOR** operation of n bits. Notations in **Algorithm 6.8** are denoted in Sect. 6.4.1.

Algorithm 6.8: N-Bits-ParallelNOR(T_0)

(1) Append-head(T_1, R_1^1).

(2) Append-head(T_2, R_1^0).

(3) $T_0 = \cup (T_1, T_2)$.

(4) **For** $k = 2$ **to** n

      (4a) Amplify(T_0, T_1, T_2).

      (4b) Append-head(T_1, R_k^1).

      (4c) Append-head(T_2, R_k^0).

      (4d) $T_0 = \cup (T_1, T_2)$.

EndFor

(5) **For** $k = 1$ **to** n

      (5a) Append-head(T_0, S_k).

EndFor

(6) **For** $k = 1$ **to** n

      (6a) **ParallelNOR(T_0, k)**.

EndFor

EndAlgorithm

Lemma 6-8: *The algorithm,* **N-Bits-ParallelNOR(T_0),** *can be used to finish the parallel* **NOR** *operation of n bits.*

Proof Steps (1) through (4d) are mainly applied to construct solution space of 2^n unsigned integers for the first input (the range of values for them is from 0 to 2^n-1). After they are finished, tube T_0 contains those inputs encoding 2^n unsigned integers. Next, Steps (5) through (5a) are used to append the value "0" or "1" for S_k (the second input) to the head of solution space of 2^n unsigned integers in tube T_0. Step (6) is the main loop and is mainly applied to finish the parallel **NOR** operation of n bits. Each execution of Step (6a) calls **ParallelNOR**(T_0, k) in Sect. 6.4.1 to finish the **NOR** operation for the kth bit of each input in 2^n inputs. Repeat execution of Step (6a) until the nth bit of each input in 2^n inputs is processed. Tube T_0 contains the result performing the parallel **NOR** operation of n bits. ∎

6.4.3 The Power for the Parallel NOR Operation of N Bits on Bio-molecular Computer

Consider that four values for an unsigned integer of two bits are, subsequently, $00(0_{10})$ $\left(R_2^0 R_1^0\right)$, $01(1_{10})$ $\left(R_2^0 R_1^1\right)$, $10(2_{10})$ $\left(R_2^1 R_1^0\right)$ and $11(3_{10})$ $\left(R_2^1 R_1^1\right)$. We want to simultaneously perform the parallel **NOR** operation for $11(3_{10})$ $\left(S_2^1 S_1^1\right)$ and those four values. **Algorithm 6.8, N-Bits-ParallelNOR**(T_0), can be employed to finish the task. Tube T_0 is an empty tube and is regarded as an input tube of **Algorithm 6.8**. From Definition 5–2, the input tube T_0 is regarded as the execution environment of the first **BMPDTM**. Similarly, tubes T_1 and T_2 used in **Algorithm 6.8** also are regarded, subsequently, as the execution environment of the second **BMPDTM** and the execution environment of the third **BMPDTM**.

Steps (1) through (4d) in **Algorithm 6.8** are applied to yield a **BMPDTM** with four bio-molecular deterministic one-tape Turing machines. After the execution for Step (1) and Step (2) of **Algorithm 6.8** is performed, tube $T_1 = \left\{R_1^1\right\}$ and tube $T_2 = \left\{R_1^0\right\}$. This is to say that a **BMDTM** in the second **BMPDTM** and in the third **BMPDTM** is constructed. Figure 6.41 is used to show the current status of the execution environment to the second **BMPDTM** and the third **BMPDTM**. From Fig. 6.41, the content of the first tape square for the tape in the first **BMDTM** in the second **BMPDTM** is written by its corresponding read-write head and is 1 ($R_1 = 1$), and the content of the first tape square for the tape in the first **BMDTM** in the third **BMPDTM** is written by its corresponding read-write head and is 0 ($R_1 = 0$). For the first **BMDTM** in the second **BMPDTM**, the position of the read-write head is moved to the *left new* tape square, and the state of the finite state control is changed as "$R_1 = 1$". Similarly, for the first **BMDTM** in the third **BMPDTM**, the position of the read-write head is also moved to the *left new* tape square, and the state of the finite state control is changed as "$R_1 = 0$".

Next, after the execution for Step (3) of **Algorithm 6.8** is implemented, tube $T_0 = \left\{R_1^1, R_1^0\right\}$, tube $T_1 = \varnothing$ and tube $T_2 = \varnothing$. This implies that the execution environment for the first bio-molecular deterministic one-tape Turing machine in the second **BMPDTM** and the first bio-molecular deterministic one-tape Turing machine in the third **BMPDTM** becomes the first **BMPDTM**. The position of the corresponding read-write head and the state of the corresponding finite state control are both reserved. Figure 6.42 is employed to explain the current status of the execution environment to the first **BMPDTM**. From Fig. 6.42, it is pointed out that the contents to the two tapes in the execution environment of the first **BMPDTM** are not changed.

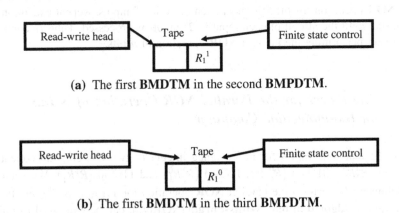

(a) The first **BMDTM** in the second **BMPDTM**.

(b) The first **BMDTM** in the third **BMPDTM**.

Fig. 6.41 Schematic representation of the current status of the execution environment to the second **BMPDTM** and the third **BMPDTM**

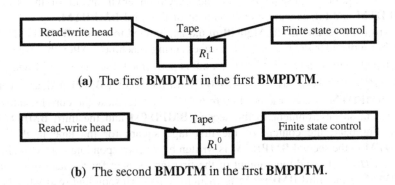

(a) The first **BMDTM** in the first **BMPDTM**.

(b) The second **BMDTM** in the first **BMPDTM**.

Fig. 6.42 Schematic representation of the current status of the execution environment to the first **BMPDTM**

Step (4) is the first loop in **Algorithm 6.8**, the upper bound (n) is two because the number of bits for representing those four values is two. Hence, after the first execution of Step (4a) is implemented, tube $T_0 = \emptyset$, tube $T_1 = \{R_1^1, R_1^0\}$ and tube $T_2 = \{R_1^1, R_1^0\}$. This indicates that the first **BMDTM** and the second **BMDTM** in the execution environment of the first **BMPDTM** are both copied into the second

BMPDTM and the third **BMPDTM**. Figure 6.43 is used to illustrate the current status of the execution environment to the second **BMPDTM** and the third **BMPDTM**. From Fig. 6.43, the contents of the first tape square for the corresponding tape of the first **BMDTM** and the corresponding tape of the second **BMDTM** in the execution environment of the second **BMPDTM** are, respectively, 1 ($R_1 = 1$) and 0 ($R_1 = 0$). The contents of the first tape square for the corresponding tape of the first **BMDTM** and the corresponding tape of the second **BMDTM** in the execution environment of the third **BMPDTM** are also, respectively, 0 ($R_1 = 0$) and 1 ($R_1 = 1$). From Fig. 6.43, four bio-molecular deterministic one-tape Turing machines are generated. For the four bio-molecular deterministic one-tape Turing machines, the position of the corresponding read-write head and the state of the corresponding finite state control are reserved.

Next, after the first execution for Step (4b) and Step (4c) in **Algorithm 6.8** is

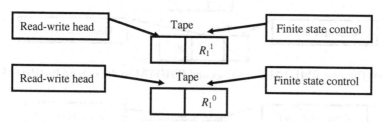

(a) The first **BMDTM** and the second **BMDTM** in the second **BMPDTM**.

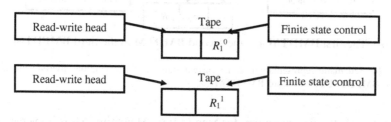

(b) The first **BMDTM** and the second **BMDTM** in the third **BMPDTM**.

Fig. 6.43 Schematic representation of the current status of the execution environment to the second **BMPDTM** and the third **BMPDTM**

implemented, tube $T_1 = \{R_2^1 R_1^1, R_2^1 R_1^0\}$ and tube $T_2 = \{R_2^0 R_1^1, R_2^0 R_1^0\}$. This is to say that the content of the *second* tape square for the tape in the first **BMDTM** in the second **BMPDTM** is written by its corresponding read-write head and is 1 ($R_2 = 1$), and the content of the *second* tape square for the tape in the second **BMDTM** in the second **BMPDTM** is written by its corresponding read-write head and is also 1 ($R_2 = 1$). Simultaneously, the position of the corresponding read-write head is both moved to the *left new* tape square and the state of the corresponding finite state control is both changed as "$R_2 = 1$". Similarly, the content of the *second* tape square for the tape in the first **BMDTM** in the third **BMPDTM** is written by its corresponding read-write head and is 0 ($R_2 = 0$), and the content of the *second* tape square for the tape in the second **BMDTM** in the third **BMPDTM** is written by its corresponding read-write head and is also 0 ($R_2 = 0$). Simultaneously, the position of the corresponding read-write head is both moved to the *left new* tape square and the state of the corresponding finite state control is both changed as "$R_2 = 0$". Figure 6.44 is applied to explain the current status of the execution environment to the second **BMPDTM** and the third **BMPDTM**.

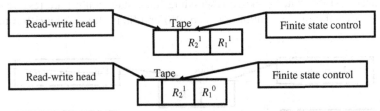

(a) The first **BMDTM** and the second **BMDTM** in the second **BMPDTM**.

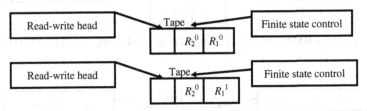

(b) The first **BMDTM** and the second **BMDTM** in the third **BMPDTM**.

Fig. 6.44 Schematic representation of the current status of the execution environment to the second **BMPDTM** and the third **BMPDTM**

Next, after the first execution for Step (4d) of **Algorithm 6.8** is implemented, tube $T_0 = \{R_2^1 R_1^1, R_2^1 R_1^0, R_2^0 R_1^1, R_2^0 R_1^0\}$, tube $T_1 = \varnothing$ and tube $T_2 = \varnothing$. This implies that the execution environment for those bio-molecular deterministic one-tape Turing machines in the second **BMPDTM** and in the third **BMPDTM** becomes the first **BMPDTM**. Figure 6.45 is employed to illustrate the current status of the execution environment to the first **BMPDTM**. From Fig. 6.45, the contents to the four tapes in the execution environment of the first **BMPDTM** are not changed, and the position of the corresponding read-write head and the state of the corresponding finite state control are reserved.

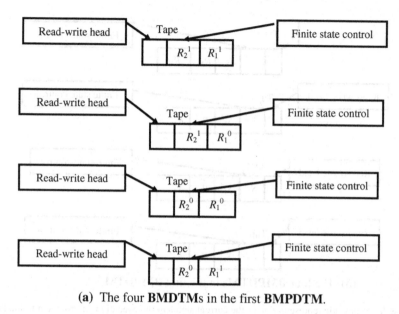

(a) The four **BMDTMs** in the first **BMPDTM**.

Fig. 6.45 Schematic representation of the current status of the execution environment to the first **BMPDTM**

Then, because Step (5) of **Algorithm 6.8** is the second loop and the upper bound for Step (5) of **Algorithm 6.8** is two, Step (5a) will be executed two times. From the first execution and the second execution of Step (5a) in **Algorithm 6.8**, S_1^1 and S_2^1 are, respectively, appended into the head of each bit pattern in tube T_0. This is to say that the contents of the third tape square and the fourth tape square for each tape of the four bio-molecular deterministic one-tape Turing machines in the first **BMPDTM** are written by the corresponding read-write head and are, subsequently, S_1^1 and S_2^1. Simultaneously, for the four bio-molecular deterministic one-tape Turing machines in the first **BMPDTM**, the position of the corresponding read-write head is all moved to the *left new* tape square, the state of the corresponding finite state control is all changed as "$S_2 = 1$" and tube $T_0 = \{S_2^1 S_1^1 R_2^1 R_1^1, S_2^1 S_1^1 R_2^1 R_1^0, S_2^1 S_1^1 R_2^0 R_1^1, S_2^1 S_1^1 R_2^0 R_1^0\}$. Figure 6.46 is used to illustrate the current status of the execution environment to the first **BMPDTM**.

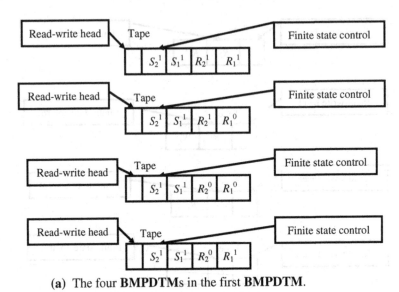

(a) The four **BMPDTMs** in the first **BMPDTM**.

Fig. 6.46 Schematic representation of the current status of the execution environment to the first **BMPDTM**

Step (6) in **Algorithm 6.8** is a third loop and is applied to carry out the parallel **NOR** operation of n bits. When the first execution of Step (6a) is implemented, it invokes **Algorithm 6.7** that is used to perform the parallel **NOR** operation of one bit, **ParallelNOR**(T_0, k), in Sect. 6.4.1. The first parameter, tube T_0, is current the execution environment of the first **BMPDTM** and contains four bio-molecular deterministic one-tape Turing machines (Fig. 6.46). It is regarded as an input tube of **Algorithm 6.7**. The value for the second parameter, k, is one and is also regarded as an input value of **Algorithm 6.7**.

When **Algorithm 6.7** is first called, seven tubes are all regarded as independent environments of seven bio-molecular parallel deterministic one-tape Turing machines. Tube T_0 is regarded as the first **BMPDTM** and tube T_k is regarded as the $(k + 1)$th **BMPDTM**. After the *first* execution of Step (1) in **Algorithm 6.7** is performed, tube $T_0 = \varnothing$, tube $T_1 = \{S_2^1 S_1^1 R_2^1 R_1^1, S_2^1 S_1^1 R_2^0 R_1^1\}$ and tube $T_2 = \{S_2^1 S_1^1 R_2^1 R_1^0, S_2^1 S_1^1 R_2^0 R_1^0\}$. This implies that the new execution environments for two bio-molecular deterministic one-tape Turing machines with the content of tape square, "R_1^1", and other two bio-molecular deterministic one-tape Turing machines with the content of tape square, "R_1^0" are, respectively, the second **BMPDTM** and the third **BMPDTM**. The position of the corresponding read-write head and the state of the corresponding finite state control are reserved. Figures 6.47 and 6.48 are used to show the result.

Then, after the *first* execution of Steps (2) and (3) in **Algorithm 6.7** is implemented, tube $T_1 = \varnothing$, tube $T_2 = \varnothing$, tube $T_4 = \varnothing$, tube $T_6 = \varnothing$, tube $T_3 = \{S_2^1 S_1^1 R_2^1 R_1^1, S_2^1 S_1^1 R_2^0 R_1^1\}$ and tube $T_5 = \{S_2^1 S_1^1 R_2^1 R_1^0, S_2^1 S_1^1 R_2^0 R_1^0\}$. This indicates that

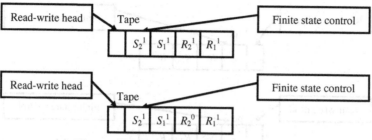

(a) The two **BMDTMs** in the second **BMPDTM**.

Fig. 6.47 Schematic representation of the current status of the execution environment to the second **BMPDTM**

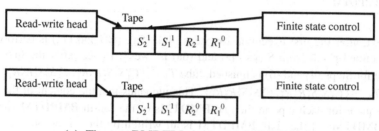

(a) The two **BMDTMs** in the third **BMPDTM**.

Fig. 6.48 Schematic representation of the current status of the execution environment to the third **BMPDTM**

the new execution environments for two bio-molecular deterministic one-tape Turing machines with the contents of tape square, "R_1^1" and "S_1^1", and other two bio-molecular deterministic one-tape Turing machines with the content of tape square, "R_1^0" and "S_1^1" are, respectively, the fourth **BMPDTM** and the sixth **BMPDTM**. The position of the corresponding read-write head and the state of the corresponding finite state control are reserved. Figures 6.49 and 6.50 are applied to explain the result.

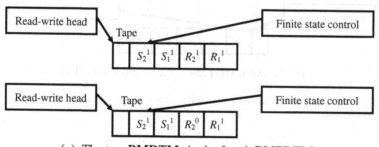

(a) The two **BMDTMs** in the fourth **BMPDTM**.

Fig. 6.49 Schematic representation of the current status of the execution environment to the fourth **BMPDTM**

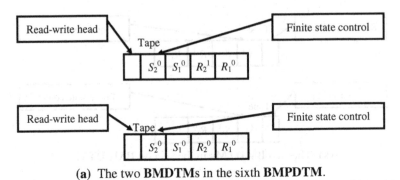

(a) The two **BMDTMs** in the sixth **BMPDTM**.

Fig. 6.50 Schematic representation of the current status of the execution environment to the sixth **BMPDTM**

Next, after the first execution for Steps (4a), (5a), (6a) and (7a) is performed, the returned result from Steps (4a) and (6a) is "yes". Hence, after the first execution for Steps (4) and (6) is finished, tube $T_3 = \{C_1^0 S_2^1 S_1^1 R_2^1 R_1^1, C_1^0 S_2^1 S_1^1 R_2^0 R_1^1\}$ and tube $T_5 = \{C_1^0 S_2^1 S_1^1 R_2^1 R_1^0, C_1^0 S_2^1 S_1^1 R_2^0 R_1^0\}$. This indicates that the content of the fifth tape square for each tape in the two **BMDTMs** of the fourth **BMPDTM** and the two **BMDTMs** of the sixth **BMPDTM** is all written by the corresponding read-write head and is 0 ($C_1 = 0$). Simultaneously, for the two **BMDTMs** of the fourth **BMPDTM** and the two **BMDTMs** of the sixth **BMPDTM**, the state of each finite state control is changed as "$C_1 = 0$" and the position of each read-write head is moved to the *left new* tape square of the corresponding tape. Figures 6.51 and 6.52 are used to subsequently show the current status of the execution environment to the fourth **BMPDTM** and the sixth **BMPDTM**.

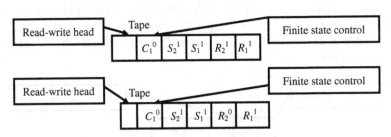

(a) The two **BMDTMs** in the fourth **BMPDTM**.

Fig. 6.51 Schematic representation of the current status of the execution environment to the fourth **BMPDTM**

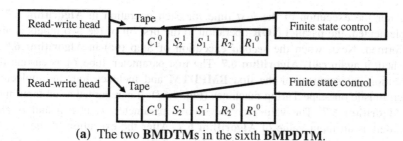

(a) The two **BMDTMs** in the sixth **BMPDTM**.

Fig. 6.52 Schematic representation of the current status of the execution environment to the sixth **BMPDTM**

Finally, after the first execution for Step (8) is implemented, tube $T_0 = \{C_1^0 S_2^1 S_1^1 R_2^1 R_1^1, C_1^0 S_2^1 S_1^1 R_2^0 R_1^1, C_1^0 S_2^1 S_1^1 R_2^1 R_1^0, C_1^0 S_2^1 S_1^1 R_2^0 R_1^0\}$, tube $T_1 = \varnothing$, tube $T_2 = \varnothing$, $T_3 = \varnothing$, $T_4 = \varnothing$, $T_5 = \varnothing$ and tube $T_6 = \varnothing$. This implies that the execution environment for the first and second bio-molecular deterministic one-tape Turing machines in the fourth **BMPDTM** and the first and second bio-molecular deterministic one-tape Turing machines in the sixth **BMPDTM** becomes the first **BMPDTM**. The position of the corresponding read-write head and the state of the corresponding finite state control are all reserved. Figure 6.53 is applied to illustrate the result.

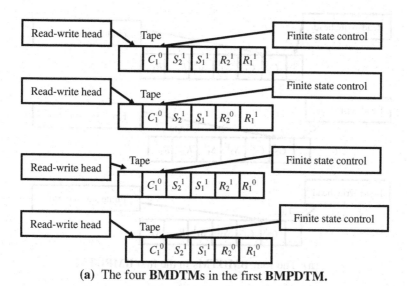

(a) The four **BMDTMs** in the first **BMPDTM**.

Fig. 6.53 Schematic representation of the current status of the execution environment to the first **BMPDTM**

After the execution of the *first* time for each operation in **Algorithm 6.7** is implemented, the parallel **NOR** operation to the first bit of the two inputs is also performed. Next, when the *second* execution of Step (6a) in **Algorithm 6.8** is finished, it again calls **Algorithm 6.7**. The first parameter, tube T_0, is current the execution environment of the first **BMPDTM** and includes four bio-molecular deterministic one-tape Turing machines (Fig. 6.53). It is regarded as an input tube of **Algorithm 6.7**. The value for the second parameter, k, is *two* and is also regarded as an input value of **Algorithm 6.7**. After the execution of the *second* time for each operation in **Algorithm 6.7** is performed, tube $T_0 = \{C_2^0 C_1^0 S_2^1 S_1^1 R_2^1 R_1^1, C_2^0 C_1^0 S_2^1 S_1^1 R_2^0 R_1^1, C_2^0 C_1^0 S_2^1 S_1^1 R_2^1 R_1^0, C_2^0 C_1^0 S_2^1 S_1^1 R_2^0 R_1^0\}$ and other tubes become all empty tubes. This indicates that the content of the sixth tape square for each **BMDTM** is written by the corresponding read-write head and is 0 ($C_2 = 0$). Simultaneously, the state of each finite state control is changed as "$C_2 = 0$" and the position of each read-write head is moved to the *left new* tape square of the corresponding tape. Figure 6.54 is used to explain the result and **Algorithm 6.8** is terminated.

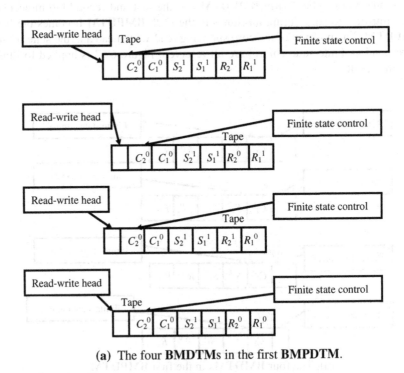

(a) The four **BMDTMs** in the first **BMPDTM**.

Fig. 6.54 Schematic representation of the current status of the execution environment to the first **BMPDTM**

6.5 Introduction for NAND Operation on Bio-molecular Computer

The **NAND** operation of a bit for two Boolean variables R and S yields a result of 0 if both R and S are 1. However, if either R or S, or both, are 0, then the result is 1. The four possible combinations for the **NAND** operation of a bit to two Boolean variables R and S are

$$\overline{0 \wedge 0} = 1$$
$$\overline{0 \wedge 1} = 1$$
$$\overline{1 \wedge 0} = 1$$
$$\overline{1 \wedge 1} = 0$$

A truth table is usually employed with logic operation to represent all possible combinations of inputs and the corresponding outputs. The truth table for the **NAND** operation is shown in Table 6.5.

Table 6.5 The truth table for the **NAND** operation of a bit for two Boolean variables R and S

Input		Output
R	S	$C = \overline{R \wedge S}$
0	0	1
0	1	1
1	0	1
1	1	0

6.5.1 The Construction for the Parallel NAND Operation of a Bit on Bio-molecular Computer

Suppose that two one-bit binary numbers, R_k and S_k, for $1 \leq k \leq n$ are employed to, respectively, represent the first input and the second input for the **NAND** operation of a bit. Also assume that a one-bit binary number, C_k, for $1 \leq k \leq n$ is applied to represent the output for the **NAND** operation of a bit. For the sake of convenience, suppose that R_k^1 denotes the fact that the value of R_k is 1 and R_k^0 denotes the fact that the value of R_k is 0. Similarly, assume that S_k^1 denotes the fact that the value of S_k is 1 and S_k^0 denotes the fact that the value of S_k is 0. Suppose that C_k^1 denotes the fact that the value of C_k is 1 and C_k^0 denotes the fact that the value of C_k is 0. The following algorithm is offered to perform the parallel **NAND** operation of a bit.

Algorithm 6.9: **ParallelNAND**(T_0, k)

   (1) $T_1 = +(T_0, R_k^1)$ and $T_2 = -(T_0, R_k^1)$.

   (2) $T_3 = +(T_1, S_k^1)$ and $T_4 = -(T_1, S_k^1)$.

   (3) $T_5 = +(T_2, S_k^1)$ and $T_6 = -(T_2, S_k^1)$.

   (4a) **If** (Detect(T_3) $= =$ "yes") **then**

      (4) Append-head(T_3, C_k^0).

  EndIf

   (5a) **If** (Detect(T_4) $= =$ "yes") **then**

      (5) Append-head(T_4, C_k^1).

  EndIf

   (6a) **If** (Detect(T_5) $= =$ "yes") **then**

      (6) Append-head(T_5, C_k^1).

  EndIf

   (7a) **If** (Detect(T_6) $= =$ "yes") **then**

      (7) Append-head(T_6, C_k^1).

  EndIf

   (8) $T_0 = \cup(T_3, T_4, T_5, T_6)$.

 EndAlgorithm

Lemma 6-9: *The algorithm,* **ParallelNAND**(T_0, k), *can be applied to perform the parallel* **NAND** *operation of a bit.*

Proof The algorithm, **ParallelNAND**(T_0, k), is implemented by means of the *extract, detect, append-head* and *merge* operations. Steps (1) through (3) employ the *extract* operations to generate different tubes including different inputs (T_1 to T_6). This indicates, T_1 contains all of the inputs that have $R_k = 1$, T_2 consists of all of the inputs that have $R_k = 0$, T_3 includes those that have $R_k = 1$ and $S_k = 1$, T_4 contains those that have $R_k = 1$ and $S_k = 0$, T_5 consists of those that have $R_k = 0$ and $S_k = 1$, and finally, T_6 includes those that have $R_k = 0$ and $S_k = 0$. Having finished Steps (1) through (3), this implies that four different inputs for the **NAND** operation of a bit as shown in Table 6.5 were poured into tubes T_3 through T_6, respectively.

Steps (4a) and (7a) are, respectively, applied to check whether includes any input for tubes T_3, T_4, T_5, and T_6 or not. If any a "yes" is returned for those steps, then the corresponding *append-head* operations will be run. Because tubes T_3, T_4, T_5 and T_6, subsequently, consists of the input for the fourth row, the third row, the second row and the first row in Table 6.5, C_k^0 from Step (4) is appended onto the head of every input in tube T_3 and C_k^1 from Steps (5) through (7) is, subsequently, appended onto the head of every input in tubes T_4, T_5, and T_6. After finishing Steps (4) through (7), four different outputs to the **NAND** operation of a bit as shown in Table 6.5 are appended into tubes T_3 through T_6. Finally, the execution of Step (8) applies the *merge* operation to pour tubes T_3 through T_6 into tube T_0. Tube T_0 contains the result performing the **NAND** operation of a bit as shown in Table 6.5. ∎

6.5.2 The Construction for the Parallel NAND Operation of N Bits on Bio-molecular Computer

The parallel **NAND** operation of n bits simultaneously produces the corresponding outputs for 2^n combinations of n bits. The following algorithm is presented to finish the parallel **NAND** operation of n bits. Notations in **Algorithm 6.10** are denoted in Sect. 6.5.1.

Algorithm 6.10: N-Bits-ParallelNAND(T_0)

(1) Append-head(T_1, $R_1{}^1$).

(2) Append-head(T_2, $R_1{}^0$).

(3) $T_0 = \cup(T_1, T_2)$.

(4) **For** $k = 2$ **to** n

    (4a) Amplify(T_0, T_1, T_2).

    (4b) Append-head(T_1, $R_k{}^1$).

    (4c) Append-head(T_2, $R_k{}^0$).

    (4d) $T_0 = \cup(T_1, T_2)$.

EndFor

(5) **For** $k = 1$ **to** n

    (5a) Append-head(T_0, S_k).

EndFor

(6) **For** $k = 1$ **to** n

    (6a) **ParallelNAND(T_0, k)**.

EndFor

EndAlgorithm

Lemma 6-10: *The algorithm,* **N-Bits-ParallelNAND(T_0),** *can be employed to perform the parallel* **NAND** *operation of n bits.*

Proof From Steps (1) through (4d), those steps are mainly used to generate solution space of 2^n combinations for the first input. After they are performed, tube T_0 includes those inputs encoding 2^n states. Next, from Steps (5) through (5a), the "append-head" operations are applied to append the value "0" or "1" for S_k (the second input) to the head of solution space of 2^n combinations in tube T_0.

Step (6) is the main loop and is mainly used to perform the parallel **NAND** operation of n bits. Each execution of Step (6a) calls **ParallelNAND(T_0, k)** in Sect. 6.5.1 to perform the **NAND** operation for the kth bit of each input in 2^n inputs. Repeat execution of Step (6a) until the nth bit of each input in 2^n inputs is processed. Tube T_0 contains the result finishing the parallel **NAND** operation of n bits. ∎

6.5.3 The Power for the Parallel NAND Operation of N Bits on Bio-molecular Computer

Consider that four values for an unsigned integer of two bits are, respectively, $00(0_{10})$ $\left(R_2^0 R_1^0\right)$, $01(1_{10})$ $\left(R_2^0 R_1^1\right)$, $10(2_{10})$ $\left(R_2^1 R_1^0\right)$ and $11(3_{10})$ $\left(R_2^1 R_1^1\right)$. We want to simultaneously carry out the parallel **NAND** operation for $00(0_{10})$ $\left(S_2^0 S_1^0\right)$ and those four values. **Algorithm 6.10**, **N-Bits-ParallelNAND(T_0)**, can be applied to perform the task. Tube T_0 is an empty tube and is regarded as an input tube of **Algorithm 6.10**. In light of Definition 5–2, the input tube T_0 is regarded as the execution environment of the first **BMPDTM**. Similarly, tubes T_1 and T_2 used in **Algorithm 6.10** also are regarded, subsequently, as the execution environment of the second **BMPDTM** and the execution environment of the third **BMPDTM**.

Steps (1) through (4d) in **Algorithm 6.10** are employed to construct a **BMPDTM** with four bio-molecular deterministic one-tape Turing machines. After the execution for Step (1) and Step (2) of **Algorithm 6.10** is finished, tube $T_1 = \{R_1^1\}$ and tube $T_1 = \{R_1^0\}$. This implies that a **BMDTM** in the second **BMPDTM** and in the third **BMPDTM** is generated. Figure 6.55 is employed to illustrate the current status of the execution environment to the second **BMPDTM** and the third **BMPDTM**. From Fig. 6.55, the content of the first tape square for the tape in the first **BMDTM** in the second **BMPDTM** is written by its corresponding read-write head and is 1 ($R_1 = 1$), and the content of the first tape square for the tape in the first **BMDTM** in the third **BMPDTM** is written by its corresponding read-write head and is 0 ($R_1 = 0$). For the first **BMDTM** in the second **BMPDTM**, the position of the read-write head is moved to the *left new* tape square, and the state of the finite state control is changed as "$R_1 = 1$". Similarly, for the first **BMDTM** in the third **BMPDTM**, the position of the read-write head is also moved to the *left new* tape square, and the state of the finite state control is changed as "$R_1 = 0$".

Next, after the execution for Step (3) of **Algorithm 6.10** is implemented, tube $T_0 = \{R_1^1, R_1^0\}$, tube $T_1 = \varnothing$ and tube $T_2 = \varnothing$. This is to say that the execution environment for the first bio-molecular deterministic one-tape Turing machine in the second **BMPDTM** and the first bio-molecular deterministic one-tape Turing machine in the third **BMPDTM** becomes the first **BMPDTM**. The position of the corresponding read-write head and the state of the corresponding finite state control are both reserved. Figure 6.56 is applied to show the current status of the execution environment to the first **BMPDTM**. From Fig. 6.56, it is indicated that the contents to the two tapes in the execution environment of the first **BMPDTM** are not changed.

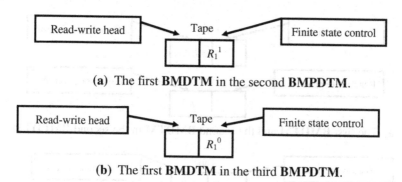

(a) The first **BMDTM** in the second **BMPDTM**.

(b) The first **BMDTM** in the third **BMPDTM**.

Fig. 6.55 Schematic representation of the current status of the execution environment to the second **BMPDTM** and the third **BMPDTM**

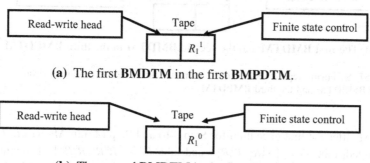

(a) The first **BMDTM** in the first **BMPDTM**.

(b) The second **BMDTM** in the first **BMPDTM**.

Fig. 6.56 Schematic representation of the current status of the execution environment to the first **BMPDTM**

Step (4) is the first loop in **Algorithm 6.10**, since the number of bits for representing those four values is two, the upper bound (n) is two. Thus, after the first execution of Step (4a) is implemented, tube $T_0 = \varnothing$, tube $T_1 = \{R_1^1, R_1^0\}$ and tube $T_2 = \{R_1^1, R_1^0\}$. This implies that the first **BMDTM** and the second **BMDTM** in the execution environment of the first **BMPDTM** are both copied into the second **BMPDTM** and the third **BMPDTM**. Figure 6.57 is employed to explain the current status of the execution environment to the second **BMPDTM** and the third **BMPDTM**. From Fig. 6.57, the contents of the first tape square for the corresponding tape of the first **BMDTM** and the corresponding tape of the second **BMDTM** in the execution environment of the second **BMPDTM** are, respectively, 1 ($R_1 = 1$) and 0 ($R_1 = 0$). The contents of the first tape square for the corresponding tape of the first **BMDTM** and the corresponding tape of the second **BMDTM** in the execution environment of the third **BMPDTM** are also, respectively, 0 ($R_1 = 0$) and 1 ($R_1 = 1$). From Fig. 6.57, four bio-molecular deterministic one-tape Turing machines are produced. For the four bio-molecular deterministic one-tape Turing machines, the position of the corresponding read-write head and the state of the corresponding finite state control are reserved.

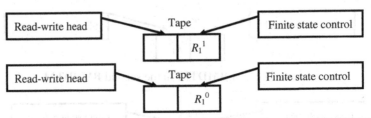

(a) The first **BMDTM** and the second **BMDTM** in the second **BMPDTM**.

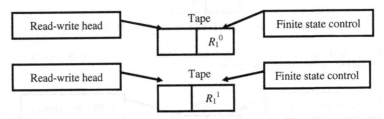

(b) The first **BMDTM** and the second **BMDTM** in the third **BMPDTM**.

Fig. 6.57 Schematic representation of the current status of the execution environment to the second **BMPDTM** and the third **BMPDTM**

Next, after the first execution for Step (4b) and Step (4c) in **Algorithm 6.10** is performed, tube $T_1 = \{R_2^1 R_1^1, R_2^1 R_1^0\}$ and tube $T_2 = \{R_2^0 R_1^1, R_2^0 R_1^0\}$. This indicates that the content of the *second* tape square for the tape in the first **BMDTM** in the second **BMPDTM** is written by its corresponding read-write head and is 1 ($R_2 = 1$), and the content of the *second* tape square for the tape in the second **BMDTM** in the second **BMPDTM** is written by its corresponding read-write head and is also 1 ($R_2 = 1$). Simultaneously, the position of the corresponding read-write head is both moved to the *left new* tape square and the state of the corresponding finite state control is both changed as "$R_2 = 1$". Similarly, the content of the *second* tape square for the tape in the first **BMDTM** in the third **BMPDTM** is written by its corresponding read-write head and is 0 ($R_2 = 0$), and the content of the *second* tape square for the tape in the second **BMDTM** in the third **BMPDTM** is written by its corresponding read-write head and is also 0 ($R_2 = 0$). Simultaneously, the position of the corresponding read-write head is both moved to the *left new* tape square and the state of the corresponding finite state control is both changed as "$R_2 = 0$". Figure 6.58 is employed to illustrate the current status of the execution environment to the second **BMPDTM** and the third **BMPDTM**.

Next, after the first execution for Step (4d) of **Algorithm 6.10** is implemented, tube $T_0 = \{R_2^1 R_1^1, R_2^1 R_1^0, R_2^0 R_1^1, R_2^0 R_1^0\}$, tube $T_1 = \varnothing$ and tube $T_2 = \varnothing$. This is to say that the execution environment for those bio-molecular deterministic one-tape Turing machines in the second **BMPDTM** and in the third **BMPDTM** becomes

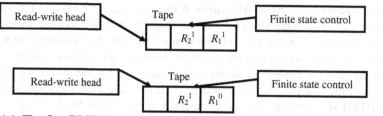

(a) The first **BMDTM** and the second **BMDTM** in the second **BMPDTM**.

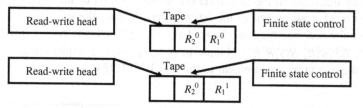

(b) The first **BMDTM** and the second **BMDTM** in the third **BMPDTM**.

Fig. 6.58 Schematic representation of the current status of the execution environment to the second **BMPDTM** and the third **BMPDTM**

the first **BMPDTM**. Figure 6.59 is used to show the current status of the execution environment to the first **BMPDTM**. From Fig. 6.59, the contents to the four tapes in the execution environment of the first **BMPDTM** are not changed, and the position of the corresponding read-write head and the state of the corresponding finite state control are reserved.

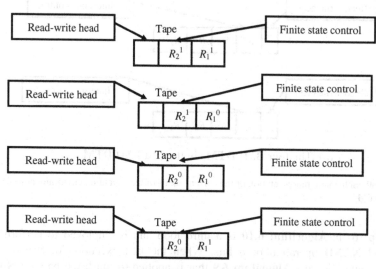

(a) The four **BMDTMs** in the first **BMPDTM**.

Fig. 6.59 Schematic representation of the current status of the execution environment to the first **BMPDTM**

Then, because Step (5) of **Algorithm 6.10** is the second loop and the upper
bound for Step (5) of **Algorithm 6.10** is two, Step (5a) will be executed two times.
From the first execution and the second execution of Step (5a) in **Algorithm 6.10**,
S_1^0 and S_2^0 are, respectively, appended into the head of each bit pattern in tube T_0.
This implies that the contents of the third tape square and the fourth tape square for
each tape of the four bio-molecular deterministic one-tape Turing machines in the
first **BMPDTM** are written by the corresponding read-write head and are, subsequently, S_1^0 and S_2^0. Simultaneously, for the four bio-molecular deterministic one-
tape Turing machines in the first **BMPDTM**, the position of the corresponding
read-write head is all moved to the *left new* tape square, the state of the corresponding finite state control is all changed as "$S_2 = 0$" and tube $T_0 = \{S_2^0 S_1^0 R_2^1 R_1^1,$
$S_2^0 S_1^0 R_2^1 R_1^0, S_2^0 S_1^0 R_2^0 R_1^1, S_2^0 S_1^0 R_2^0 R_1^0\}$. Figure 6.60 is applied to explain the current
status of the execution environment to the first **BMPDTM**.

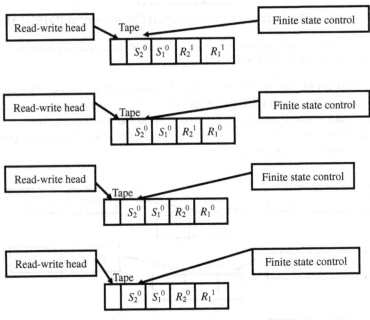

(a) The four **BMPDTMs** in the first **BMPDTM**.

Fig. 6.60 Schematic representation of the current status of the execution environment to the first
BMPDTM

Step (6) in **Algorithm 6.10** is a third loop and is employed to perform the
parallel **NAND** operation of n bits. When the first execution of Step (6a) is
implemented, it calls **Algorithm 6.9** that is applied to finish the parallel **NAND**
operation of one bit, **ParallelNAND**(T_0, k), in Sect. 6.5.1. The first parameter, tube
T_0, is current the execution environment of the first **BMPDTM** and consists of four

bio-molecular deterministic one-tape Turing machines (Fig. 6.60). It is regarded as an input tube of **Algorithm 6.9**. The value for the second parameter, k, is one and is also regarded as an input value of **Algorithm 6.9**.

When **Algorithm 6.9** is first invoked, seven tubes are all regarded as independent environments of seven bio-molecular parallel deterministic one-tape Turing machines. Tube T_0 is regarded as the first **BMPDTM** and tube T_k is regarded as the $(k+1)$th **BMPDTM**. After the *first* execution of Step (1) in **Algorithm 6.9** is finished, tube $T_0 = \varnothing$, tube $T_1 = \{S_2^0 S_1^0 R_2^1 R_1^1, S_2^0 S_1^0 R_2^0 R_1^1\}$ and tube $T_2 = \{S_2^0 S_1^0 R_2^1 R_1^0, S_2^0 S_1^0 R_2^0 R_1^0\}$. This is to say that the new execution environments for two bio-molecular deterministic one-tape Turing machines with the content of tape square, "R_1^1", and other two bio-molecular deterministic one-tape Turing machines with the content of tape square, "R_1^0" are, respectively, the second **BMPDTM** and the third **BMPDTM**. The position of the corresponding read-write head and the state of the corresponding finite state control are reserved. Figures 6.61 and 6.62 are employed to illustrate the result.

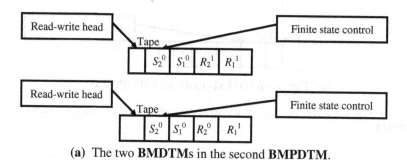

(a) The two **BMDTMs** in the second **BMPDTM**.

Fig. 6.61 Schematic representation of the current status of the execution environment to the second **BMPDTM**

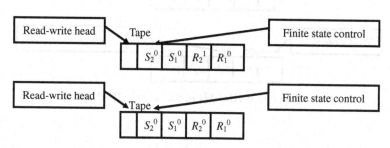

(a) The two **BMDTMs** in the third **BMPDTM**.

Fig. 6.62 Schematic representation of the current status of the execution environment to the third **BMPDTM**

Then, after the *first* execution of Steps (2) and (3) in **Algorithm 6.9** is performed, tube $T_1 = \varnothing$, tube $T_2 = \varnothing$, tube $T_3 = \varnothing$, tube $T_5 = \varnothing$, tube $T_4 = \{S_2^0 S_1^0 R_2^1 R_1^1, S_2^0 S_1^0 R_2^0 R_1^1\}$ and tube $T_6 = \{S_2^0 S_1^0 R_2^1 R_1^0, S_2^0 S_1^0 R_2^0 R_1^0\}$. This is to say that the new execution environments for two bio-molecular deterministic one-tape Turing machines with the contents of tape square, "R_1^1" and "S_1^0", and other two bio-molecular deterministic one-tape Turing machines with the content of tape square, "R_1^0" and "S_1^0" are, respectively, the fifth **BMPDTM** and the seventh **BMPDTM**. The position of the corresponding read-write head and the state of the corresponding finite state control are reserved. Figures 6.63 and 6.64 are used to show the result.

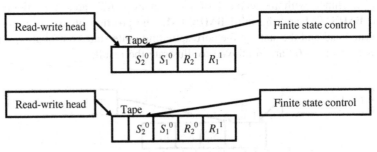

(a) The two **BMDTMs** in the fifth **BMPDTM**.

Fig. 6.63 Schematic representation of the current status of the execution environment to the fifth **BMPDTM**

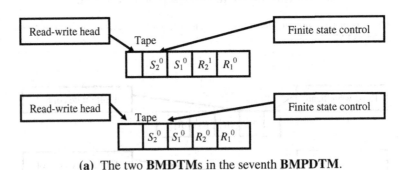

(a) The two **BMDTMs** in the seventh **BMPDTM**.

Fig. 6.64 Schematic representation of the current status of the execution environment to the seventh **BMPDTM**

Next, after the first execution for Steps (4a), (5a), (6a) and (7a) is implemented, the returned result from Steps (5a) and (7a) is "yes". Therefore, after the first execution for Steps (5) and (7) is implemented, tube $T_4 = \{C_1^1 S_2^0 S_1^0 R_2^1 R_1^1,$ $C_1^1 S_2^0 S_1^0 R_2^0 R_1^1\}$ and tube $T_6 = \{C_1^1 S_2^0 S_1^0 R_2^1 R_1^0, C_1^1 S_2^0 S_1^0 R_2^0 R_1^0\}$. This is to say that the content of the fifth tape square for each tape in the two **BMDTMs** of the fifth **BMPDTM** and the two **BMDTMs** of the seventh **BMPDTM** is all written by the corresponding read-write head and is 1 ($C_1 = 1$). Simultaneously, for the two **BMDTMs** of the fifth **BMPDTM** and the two **BMDTMs** of the seventh **BMPDTM**, the state of each finite state control is changed as "$C_1 = 1$" and the position of each read-write head is moved to the *left new* tape square of the corresponding tape. Figures 6.65 and 6.66 are applied to respectively explain the current status of the execution environment to the fifth **BMPDTM** and the seventh **BMPDTM**.

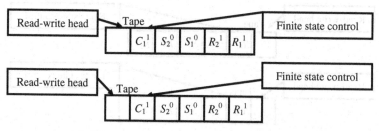

(a) The two **BMDTMs** in the fifth **BMPDTM**.

Fig. 6.65 Schematic representation of the current status of the execution environment to the fifth **BMPDTM**

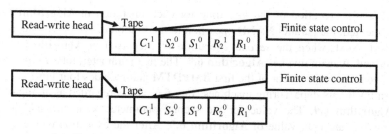

(a) The two **BMDTMs** in the seventh **BMPDTM**.

Fig. 6.66 Schematic representation of the current status of the execution environment to the seventh **BMPDTM**

Finally, after the first execution for Step (8) is performed, tube $T_0 = \{C_1^1 S_2^0 S_1^0 R_2^1 R_1^1, C_1^1 S_2^0 S_1^0 R_2^0 R_1^1, C_1^1 S_2^0 S_1^0 R_2^1 R_1^0, C_1^1 S_2^0 S_1^0 R_2^0 R_1^0\}$, tube $T_1 = \varnothing$, tube $T_2 = \varnothing$, $T_3 = \varnothing$, $T_4 = \varnothing$, $T_5 = \varnothing$ and tube $T_6 = \varnothing$. This indicates that the execution environment for the first and second bio-molecular deterministic one-tape Turing machines in the fifth **BMPDTM** and the first and second bio-molecular deterministic one-tape Turing machines in the seventh **BMPDTM** becomes the first **BMPDTM**. The position of the corresponding read-write head and the state of the corresponding finite state control are all reserved. Figure 6.67 is applied to illustrate the result.

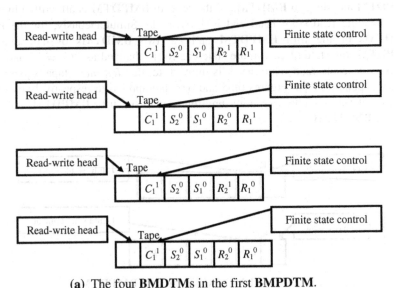

(a) The four **BMDTMs** in the first **BMPDTM**.

Fig. 6.67 Schematic representation of the current status of the execution environment to the first **BMPDTM**

After the execution of the *first* time for each operation in **Algorithm 6.9** is implemented, the parallel **NAND** operation to the first bit of the two inputs is also finished. Next, when the *second* execution of Step (6a) in **Algorithm 6.10** is performed, it again invokes **Algorithm 6.9**. The first parameter, tube T_0, is current the execution environment of the first **BMPDTM** and includes four bio-molecular deterministic one-tape Turing machines (Fig. 6.67). It is regarded as an input tube of **Algorithm 6.9**. The value for the second parameter, k, is *two* and is also regarded as an input value of **Algorithm 6.9**. After the execution of the *second* time for each operation in **Algorithm 6.9** is implemented, tube $T_0 = \{C_2^1 C_1^1 S_2^0 S_1^0 R_2^1 R_1^1, C_2^1 C_1^1 S_2^0 S_1^0 R_2^0 R_1^1, C_2^1 C_1^1 S_2^0 S_1^1 R_2^1 R_1^0, C_2^1 C_1^1 S_2^0 S_1^1 R_2^0 R_1^0\}$ and other tubes become all empty tubes. This implies that the content of the sixth tape square for each **BMDTM** is written by the corresponding read-write head and is 1 ($C_2 = 1$). Simultaneously, the state of each finite state control is changed as "$C_2 = 1$" and the position of each read-write head is moved to the *left new* tape square of the corresponding tape. Figure 6.68 is used to explain the result and **Algorithm 6.10** is terminated.

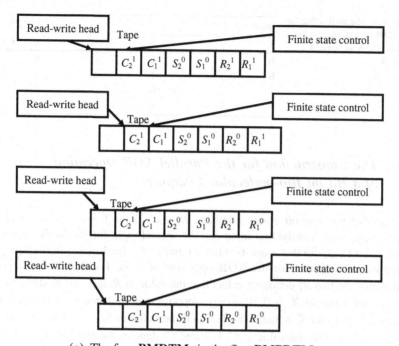

(a) The four **BMDTMs** in the first **BMPDTM**.

Fig. 6.68 Schematic representation of the current status of the execution environment to the first **BMPDTM**

6.6 Introduction for Exclusive-OR Operation on Bio-molecular Computer

The **Exclusive-OR (XOR)** operation of a bit for two Boolean variables R and S generates an output of 1 if both R and S are different values and 0 if they are the same values. The $\oplus$ symbol is applied to represent the **XOR** operation. The four possible combinations for the **XOR** operation of a bit to two Boolean variables R and S are

$$0 \oplus 0 = 0$$
$$0 \oplus 1 = 1$$
$$1 \oplus 0 = 1$$
$$1 \oplus 1 = 0$$

A truth table is usually used with logic operation to represent all possible combinations of inputs and the corresponding outputs. The truth table for the **XOR** operation is shown in Table 6.6.

Table 6.6 The truth table for the **XOR** operation of a bit for two Boolean variables R and S

Input		Output
R	S	$C = R \oplus S$
0	0	0
0	1	1
1	0	1
1	1	0

6.6.1 The Construction for the Parallel XOR Operation of a Bit on Bio-molecular Computer

Assume that two one-bit binary numbers, R_k and S_k, for $1 \leq k \leq n$ are used to, respectively, represent the first input and the second input for the **XOR** operation of a bit. Also suppose that a one-bit binary number, C_k, for $1 \leq k \leq n$ is employed to represent the output for the **XOR** operation of a bit. For the sake of convenience, assume that R_k^1 denotes the fact that the value of R_k is 1 and R_k^0 denotes the fact that the value of R_k is 0. Similarly, suppose that S_k^1 denotes the fact that the value of S_k is 1 and S_k^0 denotes the fact that the value of S_k is 0. Assume that C_k^1 denotes the fact that the value of C_k is 1 and C_k^0 denotes the fact that the value of C_k is 0. The following algorithm is proposed to perform the parallel **XOR** operation of a bit.

Algorithm 6.11: ParallelXOR(T_0, k)

(1) $T_1 = +(T_0, R_k^1)$ and $T_2 = -(T_0, R_k^1)$.

(2) $T_3 = +(T_1, S_k^1)$ and $T_4 = -(T_1, S_k^1)$.

(3) $T_5 = +(T_2, S_k^1)$ and $T_6 = -(T_2, S_k^1)$.

(4a) **If** (Detect(T_3) $= =$ "yes") **then**

    (4) Append-head(T_3, C_k^0).

EndIf

(5a) **If** (Detect(T_4) $= =$ "yes") **then**

    (5) Append-head(T_4, C_k^1).

EndIf

(6a) **If** (Detect(T_5) $= =$ "yes") **then**

    (6) Append-head(T_5, C_k^1).

EndIf

(7a) **If** (Detect(T_6) $= =$ "yes") **then**

    (7) Append-head(T_6, C_k^0).

EndIf

(8) $T_0 = \cup(T_3, T_4, T_5, T_6)$.

EndAlgorithm

Lemma 6-11: *The algorithm,* **ParallelXOR**(T_0, k), *can be employed to finish the parallel* **XOR** *operation of a bit.*

Proof The algorithm, **ParallelXOR**(T_0, k), is implemented by means of the *extract, detect, append-head* and *merge* operations. Steps (1) through (3) employ the *extract* operations to yield different tubes consisting of different inputs (T_1 to T_6). This implies, T_1 includes all of the inputs that have $R_k = 1$, T_2 contains all of the inputs that have $R_k = 0$, T_3 includes those inputs that have $R_k = 1$ and $S_k = 1$, T_4 contains those inputs that have $R_k = 1$ and $S_k = 0$, T_5 consists of those inputs that have $R_k = 0$ and $S_k = 1$, and finally, T_6 includes those that have $R_k = 0$ and $S_k = 0$. Having performed Steps (1) through (3), this is to say that four different inputs for the **XOR** operation of a bit as shown in Table 6.6 were poured into tubes T_3 through T_6, respectively.

From Steps (4a) and (7a), those steps are, respectively, applied to examine whether contains any input for tubes T_3, T_4, T_5, and T_6 or not. If any a "yes" is returned from those steps, then the corresponding *append-head* operations will be run. Because tubes T_3, T_4, T_5 and T_6, subsequently, contains the input for the fourth row, the third row, the second row and the first row in Table 6.6.1, C_k^0 from Step (4) and Step (7) is appended onto the head of every input in tubes T_3 and T_6 and C_k^1 from Steps (5) and (6) is, subsequently, appended onto the head of every input in tubes T_4 and T_5. After performing Steps (4) through (7), four different outputs to the **XOR** operation of a bit as shown in Table 6.6 are appended into tubes T_3 through T_6. Finally, the execution of Step (8) uses the *merge* operation to pour tubes T_3 through T_6 into tube T_0. Tube T_0 contains the result finishing the **XOR** operation of a bit as shown in Table 6.6. ∎

6.6.2 The Construction for the Parallel XOR Operation of N Bits on Bio-molecular Computer

The parallel **XOR** operation of n bits simultaneously generates the corresponding outputs for 2^n combinations of n bits. The following algorithm is offered to perform the parallel **XOR** operation of n bits. Notations in **Algorithm 6.12** are denoted in Sect. 6.6.1.

Algorithm 6.12: N-Bits-ParallelXOR(T_0)

(1) Append-head(T_1, R_1^1).

(2) Append-head(T_2, R_1^0).

(3) $T_0 = \cup(T_1, T_2)$.

(4) **For** $k = 2$ **to** n

    (4a) Amplify(T_0, T_1, T_2).

    (4b) Append-head(T_1, R_k^1).

    (4c) Append-head(T_2, R_k^0).

    (4d) $T_0 = \cup(T_1, T_2)$.

EndFor

(5) **For** $k = 1$ **to** n

    (5a) Append-head(T_0, S_k).

EndFor

(6) **For** $k = 1$ **to** n

    (6a) **ParallelXOR(T_0, k)**.

EndFor

EndAlgorithm

Lemma 6-12: *The algorithm, N-Bits-ParallelXOR(T_0), can be used to finish the parallel* **XOR** *operation of n bits.*

Proof Solution space of 2^n combinations for the first input is generated from Steps (1) through (4d). After they are run, tube T_0 contains those inputs encoding 2^n states. Next, from Steps (5) through (5a), the "append-head" operations are applied to append the value "0" or "1" for S_k (the second input) to the head of solution space of 2^n combinations in tube T_0.

Step (6) is the main loop and is mainly applied to run the parallel **XOR** operation of n bits. Each execution of Step (6a) calls **ParallelXOR(T_0, k)** in Sect. 6.6.1 to finish the **XOR** operation for the kth bit of each input in 2^n inputs. Repeat execution of Step (6a) until the nth bit of each input in 2^n inputs is processed. Tube T_0 includes the result performing the parallel **XOR** operation of n bits. ∎

6.6.3 The Power for the Parallel XOR Operation of N Bits on Bio-molecular Computer

Consider that four values for an unsigned integer of two bits are, respectively, $00(0_{10})$ $(R_2^0 R_1^0)$, $01(1_{10})$ $(R_2^0 R_1^1)$, $10(2_{10})$ $(R_2^1 R_1^0)$ and $11(3_{10})$ $(R_2^1 R_1^1)$. We want to simultaneously perform the parallel **XOR** operation for $10(2_{10})$ $(S_2^1 S_1^0)$ and those four values. **Algorithm 6.12, N-Bits-ParallelXOR(T_0)**, can be used to carry out the task. Tube T_0 is an empty tube and is regarded as an input tube of **Algorithm 6.12**. From Definition 5−2, the input tube T_0 is regarded as the execution environment of the first **BMPDTM**. Similarly, tubes T_1 and T_2 used in **Algorithm 6.12** also are regarded, subsequently, as the execution environment of the second **BMPDTM** and the execution environment of the third **BMPDTM**.

Steps (1) through (4d) in **Algorithm 6.12** are used to generate a **BMPDTM** with four bio-molecular deterministic one-tape Turing machines. After the execution for Step (1) and Step (2) of **Algorithm 6.12** is implemented, tube $T_1 = \{R_1^1\}$ and tube $T_2 = \{R_1^0\}$. This indicates that a **BMDTM** in the second **BMPDTM** and in the third **BMPDTM** is constructed. Figure 6.69 is applied to show the current status of the execution environment to the second **BMPDTM** and the third **BMPDTM**. From Fig. 6.69, the content of the first tape square for the tape in the first **BMDTM** in the second **BMPDTM** is written by its corresponding read-write head and is 1 ($R_1 = 1$), and the content of the first tape square for the tape in the first **BMDTM** in the third **BMPDTM** is written by its corresponding read-write head and is 0 ($R_1 = 0$). For the first **BMDTM** in the second **BMPDTM**, the position of the read-write head is moved to the *left new* tape square, and the state of the finite state control is changed as "$R_1 = 1$". Similarly, for the first **BMDTM** in the third **BMPDTM**, the position of the read-write head is also moved to the *left new* tape square, and the state of the finite state control is changed as "$R_1 = 0$".

Next, after the execution for Step (3) of **Algorithm 6.12** is performed, tube $T_0 = \{R_1^1, R_1^0\}$, tube $T_1 = \varnothing$ and tube $T_2 = \varnothing$. This implies that the execution environment for the first bio-molecular deterministic one-tape Turing machine in the second **BMPDTM** and the first bio-molecular deterministic one-tape Turing machine in the third **BMPDTM** becomes the first **BMPDTM**. The position of the corresponding read-write head and the state of the corresponding finite state control are both reserved. Figure 6.70 is employed to illustrate the current status of the execution environment to the first **BMPDTM**. From Fig. 6.70, it is pointed out that the contents to the two tapes in the execution environment of the first **BMPDTM** are not changed.

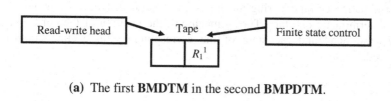

(a) The first **BMDTM** in the second **BMPDTM**.

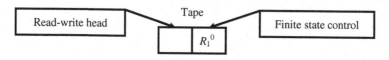

(b) The first **BMDTM** in the third **BMPDTM**.

Fig. 6.69 Schematic representation of the current status of the execution environment to the second **BMPDTM** and the third **BMPDTM**

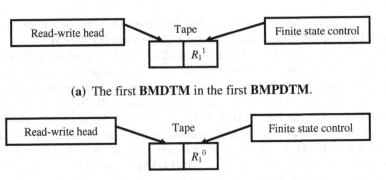

(a) The first **BMDTM** in the first **BMPDTM**.

(b) The second **BMDTM** in the first **BMPDTM**.

Fig. 6.70 Schematic representation of the current status of the execution environment to the first **BMPDTM**

Step (4) is the first loop in **Algorithm 6.12**, because the number of bits for representing those four values is two, the upper bound (n) is two. Therefore, after the first execution of Step (4a) is finished, tube $T_0 = \emptyset$, tube $T_1 = \{R_1^1, R_1^0\}$ and tube $T_2 = \{R_1^1, R_1^0\}$. This is to say that the first **BMDTM** and the second **BMDTM** in the execution environment of the first **BMPDTM** are both copied into the second **BMPDTM** and the third **BMPDTM**. Figure 6.71 is used to show the current status of the execution environment to the second **BMPDTM** and the third **BMPDTM**. From Fig. 6.71, the contents of the first tape square for the corresponding tape of the first **BMDTM** and the corresponding tape of the second **BMDTM** in the execution environment of the second **BMPDTM** are, respectively, 1 ($R_1 = 1$) and 0 ($R_1 = 0$). The contents of the first tape square for the corresponding tape of the first **BMDTM** and the corresponding tape of the second **BMDTM** in the execution environment of the third **BMPDTM** are also, respectively, 0 ($R_1 = 0$) and 1 ($R_1 = 1$). From Fig. 6.71, four bio-molecular deterministic one-tape Turing machines are generated. For the four bio-molecular deterministic one-tape Turing machines, the position of the corresponding read-write head and the state of the corresponding finite state control are reserved.

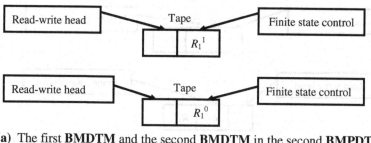

(a) The first **BMDTM** and the second **BMDTM** in the second **BMPDTM**.

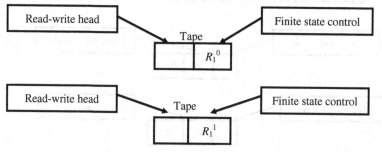

(b) The first **BMDTM** and the second **BMDTM** in the third **BMPDTM**.

Fig. 6.71 Schematic representation of the current status of the execution environment to the second **BMPDTM** and the third **BMPDTM**

Next, after the first execution for Step (4b) and Step (4c) in **Algorithm 6.12** is implemented, tube $T_1 = \{R_2^1 R_1^1, R_2^1 R_1^0\}$ and tube $T_2 = \{R_2^0 R_1^1, R_2^0 R_1^0\}$. This implies that the content of the *second* tape square for the tape in the first **BMDTM** in the second **BMPDTM** is written by its corresponding read-write head and is 1 ($R_2 = 1$), and the content of the *second* tape square for the tape in the second **BMDTM** in the second **BMPDTM** is written by its corresponding read-write head and is also 1 ($R_2 = 1$). Simultaneously, the position of the corresponding read-write head is both moved to the *left new* tape square and the state of the corresponding finite state control is both changed as "$R_2 = 1$". Similarly, the content of the *second* tape square for the tape in the first **BMDTM** in the third **BMPDTM** is written by its corresponding read-write head and is 0 ($R_2 = 0$), and the content of the *second* tape square for the tape in the second **BMDTM** in the third **BMPDTM** is written by its corresponding read-write head and is also 0 ($R_2 = 0$). Simultaneously, the position of the corresponding read-write head is both moved to the *left new* tape square and the state of the corresponding finite state control is both changed as "$R_2 = 0$". Figure 6.72 is used to explain the current status of the execution environment to the second **BMPDTM** and the third **BMPDTM**.

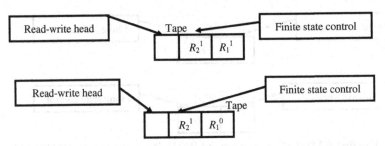

(a) The first **BMDTM** and the second **BMDTM** in the second **BMPDTM**.

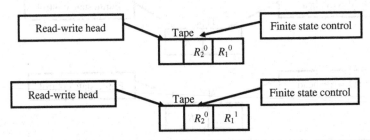

(b) The first **BMDTM** and the second **BMDTM** in the third **BMPDTM**.

Fig. 6.72 Schematic representation of the current status of the execution environment to the second **BMPDTM** and the third **BMPDTM**

Next, after the first execution for Step (4d) of **Algorithm 6.12** is finished, tube $T_0 = \{R_2^1 R_1^1, R_2^1 R_1^0, R_2^0 R_1^1, R_2^0 R_1^0\}$, tube $T_1 = \emptyset$ and tube $T_2 = \emptyset$. This indicates that the execution environment for those bio-molecular deterministic one-tape Turing machines in the second **BMPDTM** and in the third **BMPDTM** becomes the first **BMPDTM**. Figure 6.73 is employed to illustrate the current status of the execution environment to the first **BMPDTM**. From Fig. 6.73, the contents to the four tapes in the execution environment of the first **BMPDTM** are not changed, and the position of the corresponding read-write head and the state of the corresponding finite state control are reserved.

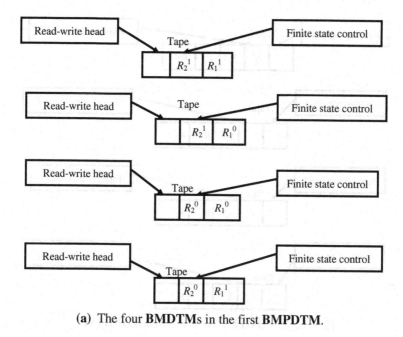

(a) The four **BMDTMs** in the first **BMPDTM**.

Fig. 6.73 Schematic representation of the current status of the execution environment to the first **BMPDTM**

Then, because Step (5) of **Algorithm 6.12** is the second loop and the upper bound for Step (5) of **Algorithm 6.12** is two, Step (5a) will be executed two times. From the first execution and the second execution of Step (5a) in **Algorithm 6.12**, S_1^0 and S_2^1 are, respectively, appended into the head of each bit pattern in tube T_0. This is to say that the contents of the third tape square and the fourth tape square for each tape of the four bio-molecular deterministic one-tape Turing machines in the first **BMPDTM** are written by the corresponding read-write head and are, subsequently, S_1^0 and S_2^1. Simultaneously, for the four bio-molecular deterministic one-tape Turing machines in the first **BMPDTM**, the position of the corresponding read-write head is all moved to the *left new* tape square, the state of the corresponding finite state control is all changed as "$S_2 = 1$" and tube $T_0 = \{ S_2^1 S_1^0 R_2^1 R_1^1, S_2^1 S_1^0 R_2^1 R_1^0, S_2^1 S_1^0 R_2^0 R_1^1, S_2^1 S_1^0 R_2^0 R_1^0 \}$. Figure 6.74 is employed to show the current status of the execution environment to the first **BMPDTM**.

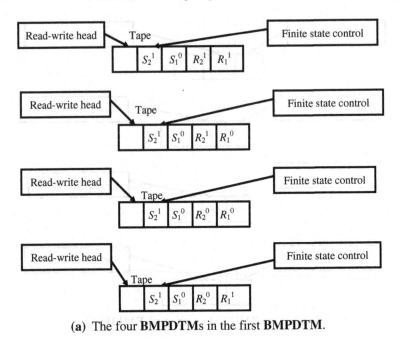

(a) The four **BMPDTMs** in the first **BMPDTM**.

Fig. 6.74 Schematic representation of the current status of the execution environment to the first **BMPDTM**

Step (6) in **Algorithm 6.12** is a third loop and is used to carry out the parallel **XOR** operation of n bits. When the first execution of Step (6a) is performed, it calls **Algorithm 6.11** that is used to perform the parallel **XOR** operation of one bit, **ParallelXOR**(T_0, k), in Sect. 6.6.1. The first parameter, tube T_0, is current the execution environment of the first **BMPDTM** and contains four bio-molecular deterministic one-tape Turing machines (Fig. 6.74). It is regarded as an input tube of **Algorithm 6.11**. The value for the second parameter, k, is one and is also regarded as an input value of **Algorithm 6.11**.

When **Algorithm 6.11** is first called, seven tubes are all regarded as independent environments of seven bio-molecular parallel deterministic one-tape Turing machines. Tube T_0 is regarded as the first **BMPDTM** and tube T_k is regarded as the $(k + 1)$th **BMPDTM**. After the *first* execution of Step (1) in **Algorithm 6.11** is performed, tube $T_0 = \varnothing$, tube $T_1 = \{S_2^1 S_1^0 R_2^1 R_1^1, S_2^1 S_1^0 R_2^0 R_1^1\}$ and tube $T_2 = \{S_2^1 S_1^0 R_2^1 R_1^0, S_2^1 S_1^0 R_2^0 R_1^0\}$. This implies that the new execution environments for two bio-molecular deterministic one-tape Turing machines with the content of tape square, "R_1^1", and other two bio-molecular deterministic one-tape Turing machines with the content of tape square, "R_1^0" are, respectively, the second **BMPDTM** and the third **BMPDTM**. The position of the corresponding read-write head and the state of the corresponding finite state control are reserved. Figures 6.75 and 6.76 are used to show the result.

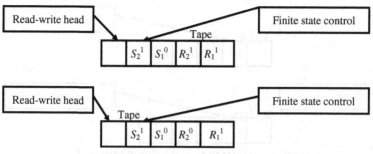

(a)   The two **BMDTMs** in the second **BMPDTM**.

Fig. 6.75 Schematic representation of the current status of the execution environment to the second **BMPDTM**

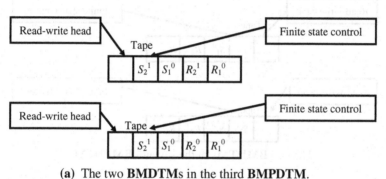

(a)   The two **BMDTMs** in the third **BMPDTM**.

Fig. 6.76 Schematic representation of the current status of the execution environment to the third **BMPDTM**

Then, after the *first* execution of Steps (2) and (3) in **Algorithm 6.11** is implemented, tube $T_1 = \varnothing$, tube $T_2 = \varnothing$, tube $T_3 = \varnothing$, tube $T_5 = \varnothing$, tube $T_4 = \{S_2^1 S_1^0 R_2^1 R_1^1, S_2^1 S_1^0 R_2^0 R_1^1\}$ and tube $T_6 = \{S_2^1 S_1^0 R_2^1 R_1^0, S_2^1 S_1^0 R_2^0 R_1^0\}$. This indicates that the new execution environments for two bio-molecular deterministic one-tape Turing machines with the contents of tape square, "R_1^1" and "S_1^0", and other two bio-molecular deterministic one-tape Turing machines with the content of tape square, "R_1^0" and "S_1^0" are, respectively, the fifth **BMPDTM** and the seventh **BMPDTM**. The position of the corresponding read-write head and the state of the corresponding finite state control are reserved. Figures 6.77 and 6.78 are employed to illustrate the result.

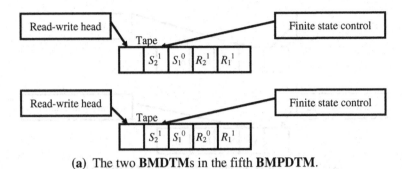

(a) The two **BMDTMs** in the fifth **BMPDTM**.

Fig. 6.77 Schematic representation of the current status of the execution environment to the fifth **BMPDTM**

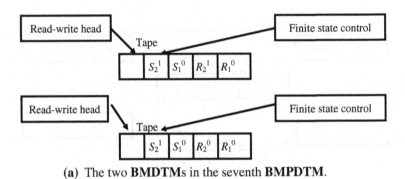

(a) The two **BMDTMs** in the seventh **BMPDTM**.

Fig. 6.78 Schematic representation of the current status of the execution environment to the seventh **BMPDTM**

Next, after the first execution for Steps (4a), (5a), (6a) and (7a) is performed, the returned result from Steps (5a) and (7a) is "yes". Thus, after the first execution for Steps (5) and (7) is finished, tube $T_4 = \left\{ C_1^1 S_2^0 S_1^0 R_2^1 R_1^1, C_1^1 S_2^0 S_1^0 R_2^0 R_1^1 \right\}$ and tube $T_6 = \left\{ C_1^0 S_2^0 S_1^0 R_2^1 R_1^0, C_1^0 S_2^0 S_1^0 R_2^0 R_1^0 \right\}$. This is to say that the content of the fifth tape square for each tape in the two **BMDTMs** of the fifth **BMPDTM** is written by the corresponding read-write head and is 1 ($C_1 = 1$). Simultaneously, the content of the fifth tape square for each tape in the two **BMDTMs** of the seventh **BMPDTM** is written by the corresponding read-write head and is 0 ($C_1 = 0$). For the two **BMDTMs** of the fifth **BMPDTM** and the two **BMDTMs** of the seventh **BMPDTM**, the state of each finite state control is, subsequently, changed as "$C_1 = 1$" and "$C_1 = 0$" and the position of each read-write head is moved to the *left new* tape square of the corresponding tape. Figures 6.79 and 6.80 are used to respectively show the current status of the execution environment to the fifth **BMPDTM** and the seventh **BMPDTM**.

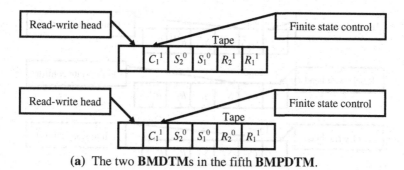

(a) The two **BMDTMs** in the fifth **BMPDTM**.

Fig. 6.79 Schematic representation of the current status of the execution environment to the fifth **BMPDTM**

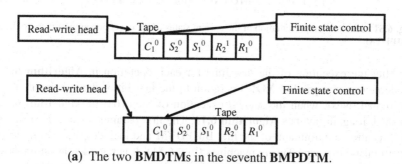

(a) The two **BMDTMs** in the seventh **BMPDTM**.

Fig. 6.80 Schematic representation of the current status of the execution environment to the seventh **BMPDTM**

Finally, after the first execution for Step (8) is implemented, tube $T_0 = \{C_1^1 S_2^0 S_1^0 R_2^1 R_1^1, C_1^1 S_2^0 S_1^0 R_2^0 R_1^1, C_1^0 S_2^0 S_1^0 R_2^1 R_1^0, C_1^0 S_2^0 S_1^0 R_2^0 R_1^1\}$, tube $T_1 = \varnothing$, tube $T_2 = \varnothing$, $T_3 = \varnothing$, $T_4 = \varnothing$, $T_5 = \varnothing$ and tube $T_6 = \varnothing$. This is to say that the execution environment for the first and second bio-molecular deterministic one-tape Turing machines in the fifth **BMPDTM** and the first and second bio-molecular deterministic one-tape Turing machines in the seventh **BMPDTM** becomes the first **BMPDTM**. The position of the corresponding read-write head and the state of the corresponding finite state control are all reserved. Figure 6.81 is employed to explain the result.

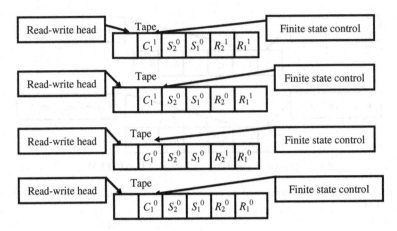

(a) The four **BMDTMs** in the first **BMPDTM**.

Fig. 6.81 Schematic representation of the current status of the execution environment to the first **BMPDTM**

After the execution of the *first* time for each operation in **Algorithm 6.11** is implemented, the parallel **XOR** operation to the first bit of the two inputs is also performed. Next, when the *second* execution of Step (6a) in **Algorithm 6.12** is finished, it again invokes **Algorithm 6.11**. The first parameter, tube T_0, is current the execution environment of the first **BMPDTM** and includes four bio-molecular deterministic one-tape Turing machines (Fig. 6.81). It is regarded as an input tube of **Algorithm 6.11**. The value for the second parameter, k, is *two* and is also regarded as an input value of **Algorithm 6.11**. After the execution of the *second* time for each operation in **Algorithm 6.11** is implemented, tube $T_0 = \{C_2^1 C_1^1 S_2^0 S_1^0 R_2^1 R_1^1, C_2^0 C_1^1 S_2^0 S_1^0 R_2^0 R_1^1, C_2^1 C_1^0 S_2^0 S_1^0 R_2^1 R_1^0, C_2^0 C_1^0 S_2^0 S_1^0 R_2^0 R_1^0\}$ and other tubes become all empty tubes. This indicates that the content of the sixth tape square for each **BMDTM** is written by the corresponding read-write head and is, subsequently, 1 ($C_2 = 1$), 0 ($C_2 = 0$), 1 ($C_2 = 1$) and 0 ($C_2 = 0$). Simultaneously, the state of each finite state control is, respectively, changed as "$C_2 = 1$", "$C_2 = 0$", "$C_2 = 1$" and "$C_2 = 0$" and the position of each read-write head is moved to the *left new* tape square of the corresponding tape. Figure 6.82 is used to explain the result and **Algorithm 6.12** is terminated.

6.7 Introduction for Exclusive-NOR Operation on Bio-molecular Computer

The one's complement of the **Exclusive-OR (XOR)** operation of a bit for two Boolean variables R and S is known as the **Exclusive-NOR (XNOR)** operation. The **Exclusive-NOR (XNOR)** operation of a bit for two Boolean variables R and S

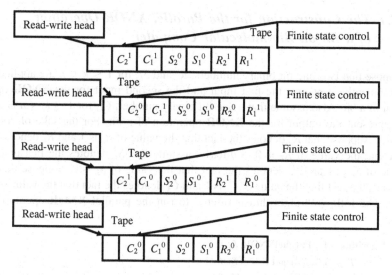

(a) The four **BMDTMs** in the first **BMPDTM**.

Fig. 6.82 Schematic representation of the current status of the execution environment to the first **BMPDTM**

generates an output of 1 if both R and S are the same values and 0 if they are different values. The four possible combinations for the **XNOR** operation of a bit to two Boolean variables R and S are

$$\overline{0 \oplus 0} = 1$$
$$\overline{0 \oplus 1} = 0$$
$$\overline{1 \oplus 0} = 0$$
$$\overline{1 \oplus 1} = 1$$

A truth table is usually employed with logic operation to represent all possible combinations of inputs and the corresponding outputs. The truth table for the **XNOR** operation is shown in Table 6.7.

Table 6.7 The truth table for the **XNOR** operation of a bit for two Boolean variables R and S

Input		Output
R	S	$C = \overline{R \oplus S}$
0	0	1
0	1	0
1	0	0
1	1	1

6.7.1 The Construction for the Parallel XNOR Operation of a Bit on Bio-molecular Computer

Suppose that two one-bit binary numbers, R_k and S_k, for $1 \leq k \leq n$ are applied to, subsequently, represent the first input and the second input for the **XNOR** operation of a bit. Also assume that a one-bit binary number, C_k, for $1 \leq k \leq n$ is used to represent the output for the **XNOR** operation of a bit. For the sake of convenience, suppose that R_k^1 denotes the fact that the value of R_k is 1 and R_k^0 denotes the fact that the value of R_k is 0. Similarly, assume that S_k^1 denotes the fact that the value of S_k is 1 and S_k^0 denotes the fact that the value of S_k is 0. Suppose that C_k^1 denotes the fact that the value of C_k is 1 and C_k^0 denotes the fact that the value of C_k is 0. The following algorithm is offered to run the parallel **XNOR** operation of a bit.

Algorithm 6.13: ParallelXNOR(T_0, k)

(1) $T_1 = +(T_0, R_k^1)$ and $T_2 = -(T_0, R_k^1)$.

(2) $T_3 = +(T_1, S_k^1)$ and $T_4 = -(T_1, S_k^1)$.

(3) $T_5 = +(T_2, S_k^1)$ and $T_6 = -(T_2, S_k^1)$.

(4a) **If** (Detect(T_3) $==$ "yes") **then**

    (4) Append-head(T_3, C_k^1).

EndIf

(5a) **If** (Detect(T_4) $==$ "yes") **then**

    (5) Append-head(T_4, C_k^0).

EndIf

(6a) **If** (Detect(T_5) $==$ "yes") **then**

    (6) Append-head(T_5, C_k^0).

EndIf

(7a) **If** (Detect(T_6) $==$ "yes") **then**

    (7) Append-head(T_6, C_k^1).

EndIf

(8) $T_0 = \cup(T_3, T_4, T_5, T_6)$.

EndAlgorithm

Lemma 6-13: *The algorithm,* **ParallelXNOR(T_0, k),** *can be used to perform the parallel* **XNOR** *operation of a bit.*

Proof The algorithm, **ParallelXNOR(T_0, k)**, is implemented by means of the *extract*, *detect*, *append-head* and *merge* operations. Steps (1) through (3) employ the *extract* operations to produce different tubes including different inputs (T_1 to T_6). This is to say that T_1 contains all of the inputs that have $R_k = 1$, T_2 consists of all of the inputs that have $R_k = 0$, T_3 includes those inputs that have $R_k = 1$ and $S_k = 1$, T_4 contains those inputs that have $R_k = 1$ and $S_k = 0$, T_5 consists of those inputs that have $R_k = 0$ and $S_k = 1$, and finally, T_6 includes those that have $R_k = 0$ and $S_k = 0$. Having finished Steps (1) through (3), this implies that four different inputs for the **XNOR** operation of a bit as shown in Table 6.7 were poured into tubes T_3 through T_6, respectively.

From Steps (4a) and (7a), those steps are, subsequently, used to test whether includes any input for tubes T_3, T_4, T_5, and T_6 or not. If any a "yes" is returned from those steps, then the corresponding *append-head* operations will be run. Because tubes T_3, T_4, T_5 and T_6, subsequently, consists of the input for the fourth row, the third row, the second row and the first row in Table 6.7, C_k^1 from Step (4) and Step (7) is appended onto the head of every input in tubes T_3 and T_6 and C_k^0 from Steps (5) and (6) is, subsequently, appended onto the head of every input in tubes T_4 and T_5. After performing Steps (4) through (7), four different outputs to the **XNOR** operation of a bit as shown in Table 6.7 are appended into tubes T_3 through T_6. Finally, the execution of Step (8) applies the *merge* operation to pour tubes T_3 through T_6 into tube T_0. Tube T_0 contains the result finishing the **XNOR** operation of a bit as shown in Table 6.7. ∎

6.7.2 The Construction for the Parallel XNOR Operation of N Bits on Bio-molecular Computer

The parallel **XNOR** operation of n bits simultaneously produces the corresponding outputs for 2^n combinations of n bits. The following algorithm is proposed to finish the parallel **XNOR** operation of n bits. Notations in **Algorithm 6.14** are denoted in Sect. 6.7.1.

Algorithm 6.14: N-Bits-ParallelXNOR(T_0)
(1) Append-head(T_1, R_1^1).
(2) Append-head(T_2, R_1^0).
(3) $T_0 = \cup(T_1, T_2)$.
(4) **For** $k = 2$ **to** n
    (4a) Amplify(T_0, T_1, T_2).
    (4b) Append-head(T_1, R_k^1).
    (4c) Append-head(T_2, R_k^0).
    (4d) $T_0 = \cup(T_1, T_2)$.
EndFor
(5) **For** $k = 1$ **to** n
    (5a) Append-head(T_0, S_k).
EndFor
(6) **For** $k = 1$ **to** n
    (6a) **ParallelXNOR(T_0, k)**.
EndFor
EndAlgorithm

Lemma 6-14: *The algorithm,* **N-Bits-ParallelXNOR**(T_0), *can be applied to perform the parallel* **XNOR** *operation of n bits.*

Proof Steps (1) through (4d) are mainly used to yield solution space of 2^n combinations for the first input. After they are performed, tube T_0 contains those inputs encoding 2^n states. Next, Steps (5) through (5a) use the "append-head" operations to append the value "0" or "1" for S_k (the second input) to the head of solution space of 2^n combinations in tube T_0.

Step (6) is the main loop and is mainly employed to finish the parallel **XNOR** operation of n bits. Each execution of Step (6a) calls **ParallelXNOR**(T_0, k) in Sect. 6.7.1 to perform the **XNOR** operation for the kth bit of each input in 2^n inputs. Repeat execution of Step (6a) until the nth bit of each input in 2^n inputs is processed. Tube T_0 includes the result finishing the parallel **XNOR** operation of n bits. ∎

6.7.3 The Power for the Parallel XNOR Operation of N Bits on Bio-molecular Computer

Consider that four values for an unsigned integer of two bits are, respectively, $00(0_{10})$ $(R_2^0 R_1^0)$, $01(1_{10})$ $(R_2^0 R_1^1)$, $10(2_{10})$ $(R_2^1 R_1^0)$ and $11(3_{10})$ $(R_2^1 R_1^1)$. We want to simultaneously carry out the parallel **XNOR** operation for $01(1_{10})$ $(S_2^0 S_1^1)$ and those four values. **Algorithm 6.14**, **N-Bits-ParallelXNOR**(T_0), can be applied to perform the task. Tube T_0 is an empty tube and is regarded as an input tube of **Algorithm 6.14**. According to Definition 5–2, the input tube T_0 is regarded as the execution environment of the first **BMPDTM**. Similarly, tubes T_1 and T_2 used in **Algorithm 6.14** also are regarded, subsequently, as the execution environment of the second **BMPDTM** and the execution environment of the third **BMPDTM**.

Steps (1) through (4d) in **Algorithm 6.14** are used to construct a **BMPDTM** with four bio-molecular deterministic one-tape Turing machines. After the execution for Step (1) and Step (2) of **Algorithm 6.14** is performed, tube $T_1 = \{R_1^1\}$ and tube $T_2 = \{R_1^0\}$. This implies that a **BMDTM** in the second **BMPDTM** and in the third **BMPDTM** is generated. Figure 6.83 is used to explain the current status of the execution environment to the second **BMPDTM** and the third **BMPDTM**. From Fig. 6.83, the content of the first tape square for the tape in the first **BMDTM** in the second **BMPDTM** is written by its corresponding read-write head and is 1 ($R_1 = 1$), and the content of the first tape square for the tape in the first **BMDTM** in the third **BMPDTM** is written by its corresponding read-write head and is 0 ($R_1 = 0$). For the first **BMDTM** in the second **BMPDTM**, the position of the read-write head is moved to the *left new* tape square, and the state of the finite state control is changed as "$R_1 = 1$". Similarly, for the first **BMDTM** in the third **BMPDTM**, the position of the read-write head is also moved to the *left new* tape square, and the state of the finite state control is changed as "$R_1 = 0$".

Next, after the execution for Step (3) of **Algorithm 6.14** is finished, tube $T_0 = \{R_1^1, R_1^0\}$, tube $T_1 = \varnothing$ and tube $T_2 = \varnothing$. This indicates that the execution environment for the first bio-molecular deterministic one-tape Turing machine in the second **BMPDTM** and the first bio-molecular deterministic one-tape Turing machine in the third **BMPDTM** becomes the first **BMPDTM**. The position of the corresponding read-write head and the state of the corresponding finite state control are both reserved. Figure 6.84 is used to show the current status of the execution environment to the first **BMPDTM**. From Fig. 6.84, it is indicated that the contents to the two tapes in the execution environment of the first **BMPDTM** are not changed.

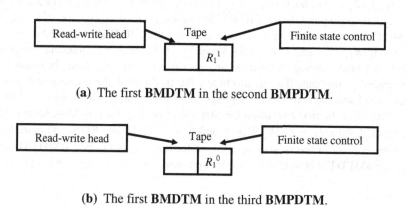

(a) The first **BMDTM** in the second **BMPDTM**.

(b) The first **BMDTM** in the third **BMPDTM**.

Fig. 6.83 Schematic representation of the current status of the execution environment to the second **BMPDTM** and the third **BMPDTM**

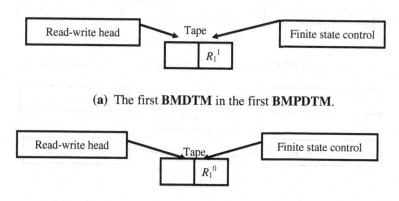

(a) The first **BMDTM** in the first **BMPDTM**.

(b) The second **BMDTM** in the first **BMPDTM**.

Fig. 6.84 Schematic representation of the current status of the execution environment to the first **BMPDTM**

Step (4) is the first loop in **Algorithm 6.14**, since the number of bits for representing those four values is two, the upper bound (n) is two. Hence, after the first execution of Step (4a) is performed, tube $T_0 = \varnothing$, tube $T_1 = \{R_1^1, R_1^0\}$ and tube $T_2 = \{R_1^1, R_1^0\}$. This is to say that the first **BMDTM** and the second **BMDTM** in the execution environment of the first **BMPDTM** are both copied into the second **BMPDTM** and the third **BMPDTM**. Figure 6.85 is applied to illustrate the current status of the execution environment to the second **BMPDTM** and the third **BMPDTM**. From Fig. 6.85, the contents of the first tape square for the corresponding tape of the first **BMDTM** and the corresponding tape of the second **BMDTM** in the execution environment of the second **BMPDTM** are, respectively, 1 ($R_1 = 1$) and 0 ($R_1 = 0$). The contents of the first tape square for the corresponding tape of the first **BMDTM** and the corresponding tape of the second **BMDTM** in the execution environment of the third **BMPDTM** are also, respectively, 0 ($R_1 = 0$) and 1 ($R_1 = 1$). From Fig. 6.85, four bio-molecular deterministic one-tape Turing machines are produced. For the four bio-molecular deterministic one-tape Turing machines, the position of the corresponding read-write head and the state of the corresponding finite state control are reserved.

Next, after the first execution for Step (4b) and Step (4c) in **Algorithm 6.14** is performed, tube $T_1 = \{R_2^1 R_1^1, R_2^1 R_1^0\}$ and tube $T_2 = \{R_2^0 R_1^1, R_2^0 R_1^0\}$. This indicates that the content of the *second* tape square for the tape in the first **BMDTM** in the second **BMPDTM** is written by its corresponding read-write head and is 1 ($R_2 = 1$),

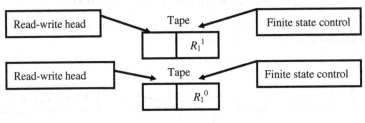

(a) The first **BMDTM** and the second **BMDTM** in the second **BMPDTM**.

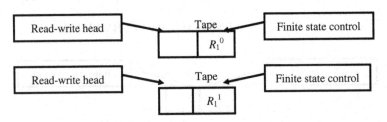

(b) The first **BMDTM** and the second **BMDTM** in the third **BMPDTM**.

Fig. 6.85 Schematic representation of the current status of the execution environment to the second **BMPDTM** and the third **BMPDTM**

and the content of the *second* tape square for the tape in the second **BMDTM** in the second **BMPDTM** is written by its corresponding read-write head and is also 1 ($R_2 = 1$). Simultaneously, the position of the corresponding read-write head is both moved to the *left new* tape square and the state of the corresponding finite state control is both changed as "$R_2 = 1$". Similarly, the content of the *second* tape square for the tape in the first **BMDTM** in the third **BMPDTM** is written by its corresponding read-write head and is 0 ($R_2 = 0$), and the content of the *second* tape square for the tape in the second **BMDTM** in the third **BMPDTM** is written by its corresponding read-write head and is also 0 ($R_2 = 0$). Simultaneously, the position of the corresponding read-write head is both moved to the *left new* tape square and the state of the corresponding finite state control is both changed as "$R_2 = 0$". Figure 6.86 is applied to show the current status of the execution environment to the second **BMPDTM** and the third **BMPDTM**.

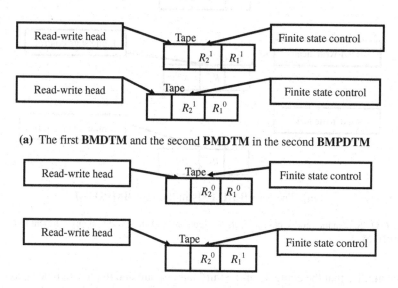

(a) The first **BMDTM** and the second **BMDTM** in the second **BMPDTM**

(b) The first **BMDTM** and the second **BMDTM** in the third **BMPDTM**

Fig. 6.86 Schematic representation of the current status of the execution environment to the second **BMPDTM** and the third **BMPDTM**

Next, after the first execution for Step (4d) of **Algorithm 6.14** is implemented, tube $T_0 = \{R_2^1 R_1^1, R_2^1 R_1^0, R_2^0 R_1^1, R_2^0 R_1^0\}$, tube $T_1 = \varnothing$ and tube $T_2 = \varnothing$. This is to say that the execution environment for those bio-molecular deterministic one-tape Turing machines in the second **BMPDTM** and in the third **BMPDTM** becomes the first **BMPDTM**. Figure 6.87 is used to explain the current status of the execution environment to the first **BMPDTM**. From Fig. 6.87, the contents to the four tapes in the execution environment of the first **BMPDTM** are not changed, and the position of the corresponding read-write head and the state of the corresponding finite state control are reserved.

Then, because Step (5) of **Algorithm 6.14** is the second loop and the upper bound for Step (5) of **Algorithm 6.14** is two, Step (5a) will be executed two times. From the first execution and the second execution of Step (5a) in **Algorithm 6.14**, S_1^1 and S_2^0 are, respectively, appended into the head of each bit pattern in tube T_0.

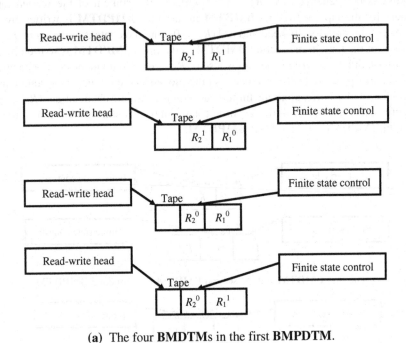

(a) The four **BMDTMs** in the first **BMPDTM**.

Fig. 6.87 Schematic representation of the current status of the execution environment to the first **BMPDTM**

This implies that the contents of the third tape square and the fourth tape square for each tape of the four bio-molecular deterministic one-tape Turing machines in the first **BMPDTM** are written by the corresponding read-write head and are, subsequently, S_1^1 and S_2^0. Simultaneously, for the four bio-molecular deterministic one-tape Turing machines in the first **BMPDTM**, the position of the corresponding read-write head is all moved to the *left new* tape square, the state of the corresponding finite state control is all changed as "$S_2 = 0$" and tube $T_0 = \{S_2^0 S_1^1 R_2^1 R_1^1, S_2^0 S_1^1 R_2^1 R_1^0, S_2^0 S_1^1 R_2^0 R_1^1, S_2^0 S_1^1 R_2^0 R_1^0\}$. Figure 6.88 is applied to illustrate the current status of the execution environment to the first **BMPDTM**.

Step (6) in **Algorithm 6.14** is a third loop and is employed to perform the parallel **XNOR** operation of n bits. When the first execution of Step (6a) is finished, it calls **Algorithm 6.13** that is applied to carry out the parallel **XNOR**

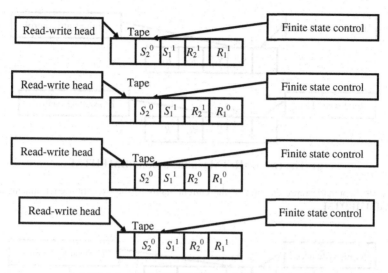

(a) The four **BMPDTMs** in the first **BMPDTM**.

Fig. 6.88 Schematic representation of the current status of the execution environment to the first **BMPDTM**

operation of one bit, **ParallelXNOR**(T_0, k), in Sect. 6.7.1. The first parameter, tube T_0, is current the execution environment of the first **BMPDTM** and includes four bio-molecular deterministic one-tape Turing machines (Fig. 6.88). It is regarded as an input tube of **Algorithm 6.13**. The value for the second parameter, k, is one and is also regarded as an input value of **Algorithm 6.13**.

When **Algorithm 6.13** is first invoked, seven tubes are all regarded as independent environments of seven bio-molecular parallel deterministic one-tape Turing machines. Tube T_0 is regarded as the first **BMPDTM** and tube T_k is regarded as the $(k + 1)$th **BMPDTM**. After the *first* execution of Step (1) in **Algorithm 6.13** is performed, tube $T_0 = \varnothing$, tube $T_1 = \left\{ S_2^0 S_1^1 R_2^1 R_1^1, S_2^0 S_1^1 R_2^0 R_1^1 \right\}$ and tube $T_2 = \left\{ S_2^0 S_1^1 R_2^1 R_1^0, S_2^0 S_1^1 R_2^0 R_1^0 \right\}$. This is to say that the new execution environments for two bio-molecular deterministic one-tape Turing machines with the content of tape square, "R_1^1", and other two bio-molecular deterministic one-tape Turing machines with the content of tape square, "R_1^0" are, respectively, the second **BMPDTM** and the third **BMPDTM**. The position of the corresponding read-write head and the state of the corresponding finite state control are reserved. Figures 6.89 and 6.90 are applied to explain the result.

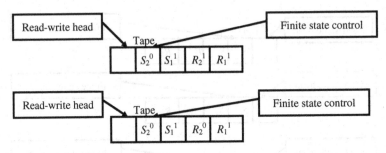

(a) The two **BMDTMs** in the second **BMPDTM**.

Fig. 6.89 Schematic representation of the current status of the execution environment to the second **BMPDTM**

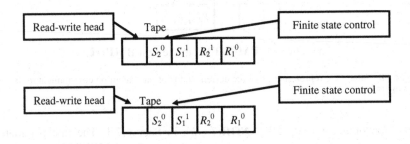

(a) The two **BMDTMs** in the third **BMPDTM**.

Fig. 6.90 Schematic representation of the current status of the execution environment to the third **BMPDTM**

Then, after the *first* execution of Steps (2) and (3) in **Algorithm 6.13** is finished, tube $T_1 = \varnothing$, tube $T_2 = \varnothing$, tube $T_4 = \varnothing$, tube $T_6 = \varnothing$, tube $T_3 = \{S_2^0 S_1^1 R_2^1 R_1^1,$ $S_2^0 S_1^1 R_2^0 R_1^1\}$ and tube $T_5 = \{S_2^0 S_1^1 R_2^1 R_1^0, S_2^0 S_1^1 R_2^0 R_1^0\}$. This indicates that the new execution environments for two bio-molecular deterministic one-tape Turing machines with the contents of tape square, "R_1^1" and "S_1^1", and other two bio-molecular deterministic one-tape Turing machines with the content of tape square, "R_1^0" and "S_1^1" are, respectively, the fourth **BMPDTM** and the sixth **BMPDTM**. The position of the corresponding read-write head and the state of the corresponding finite state control are reserved. Figures 6.91 and 6.92 are used to show the result.

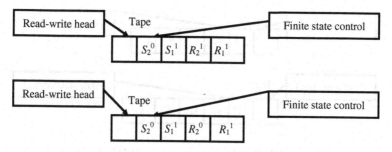

(a) The two **BMDTM**s in the fourth **BMPDTM**.

Fig. 6.91 Schematic representation of the current status of the execution environment to the fourth **BMPDTM**

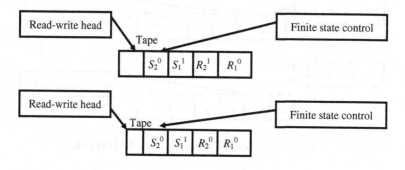

(a) The two **BMDTM**s in the sixth **BMPDTM**.

Fig. 6.92 Schematic representation of the current status of the execution environment to the sixth **BMPDTM**

Next, after the first execution for Steps (4a), (5a), (6a) and (7a) is implemented, the returned result from Steps (4a) and (6a) is "yes". Therefore, after the first execution for Steps (4) and (6) is performed, tube $T_3 = \{C_1^1 S_2^0 S_1^1 R_2^1 R_1^1, C_1^1 S_2^0 S_1^1 R_2^0 R_1^1\}$ and tube $T_5 = \{C_1^0 S_2^0 S_1^1 R_2^1 R_1^0, C_1^0 S_2^0 S_1^1 R_2^0 R_1^0\}$. This indicates that the content of the fifth tape square for each tape in the two **BMDTM**s of the fourth **BMPDTM** is written by the corresponding read-write head and is 1 ($C_1 = 1$). Simultaneously, the content of the fifth tape square for each tape in the two **BMDTM**s of the sixth **BMPDTM** is written by the corresponding read-write head and is 0 ($C_1 = 0$). For the two **BMDTM**s of the fourth **BMPDTM** and the two **BMDTM**s of the sixth **BMPDTM**, the state of each finite state control is, subsequently, changed as "$C_1 = 1$" and "$C_1 = 0$" and the position of each read-write head is moved to the *left new* tape square of the corresponding tape. Figures 6.93 and 6.94 are employed to respectively illustrate the current status of the execution environment to the fourth **BMPDTM** and the sixth **BMPDTM**.

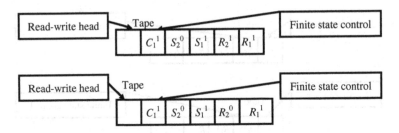

(a) The two **BMDTM**s in the fourth **BMPDTM**.

Fig. 6.93 Schematic representation of the current status of the execution environment to the fourth **BMPDTM**

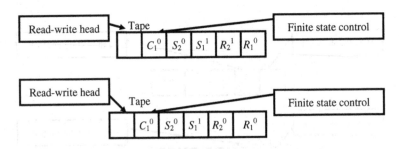

(a) The two **BMDTM**s in the sixth **BMPDTM**.

Fig. 6.94 Schematic representation of the current status of the execution environment to the sixth **BMPDTM**

Finally, after the first execution for Step (8) is implemented, tube $T_0 = \{C_1^1 S_2^0 S_1^1 R_2^1 R_1^1, C_1^1 S_2^0 S_1^1 R_2^0 R_1^1, C_1^0 S_2^0 S_1^1 R_2^1 R_1^0, C_1^0 S_2^0 S_1^1 R_2^0 R_1^0\}$, tube $T_1 = \varnothing$, tube $T_2 = \varnothing$, $T_3 = \varnothing$, $T_4 = \varnothing$, $T_5 = \varnothing$ and tube $T_6 = \varnothing$. This implies that the execution environment for the first and second bio-molecular deterministic one-tape Turing machines in the fourth **BMPDTM** and the first and second bio-molecular deterministic one-tape Turing machines in the sixth **BMPDTM** becomes the first **BMPDTM**. The position of the corresponding read-write head and the state of the corresponding finite state control are all reserved. Figure 6.95 is applied to show the result.

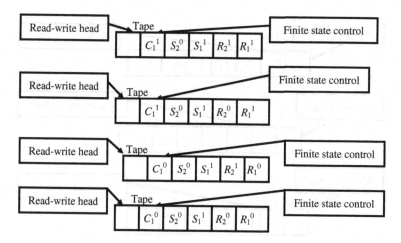

(a) The four **BMDTMs** in the first **BMPDTM**.

Fig. 6.95 Schematic representation of the current status of the execution environment to the first **BMPDTM**

After the execution of the *first* time for each operation in **Algorithm 6.13** is performed, the parallel **XNOR** operation to the first bit of the two inputs is also carried out. Next, when the *second* execution of Step (6a) in **Algorithm 6.14** is finished, it again invokes **Algorithm 6.13**. The first parameter, tube T_0, is current the execution environment of the first **BMPDTM** and includes four bio-molecular deterministic one-tape Turing machines (Fig. 6.95). It is regarded as an input tube of **Algorithm 6.13**. The value for the second parameter, k, is *two* and is also regarded as an input value of **Algorithm 6.13**. After the execution of the *second* time for each operation in **Algorithm 6.13** is performed, tube $T_0 = \{C_2^0 C_1^1 S_2^0 S_1^1 R_2^1 R_1^1, C_2^1 C_1^1 S_2^0 S_1^1 R_2^0 R_1^1, C_2^0 C_1^0 S_2^0 S_1^1 R_2^1 R_1^0, C_2^1 C_1^0 S_2^0 S_1^1 R_2^0 R_1^0\}$ and other tubes become all empty tubes. This indicates that the content of the sixth tape square for each **BMDTM** is written by the corresponding read-write head and is, subsequently, 0 ($C_2 = 0$), 1 ($C_2 = 1$), 0 ($C_2 = 0$) and 1 ($C_2 = 1$). Simultaneously, the state of each finite state control is, respectively, changed as "$C_2 = 0$", "$C_2 = 1$", "$C_2 = 0$" and "$C_2 = 1$" and the position of each read-write head is moved to the *left new* tape square of the corresponding tape. Figure 6.96 is applied to illustrate the result and **Algorithm 6.14** is terminated.

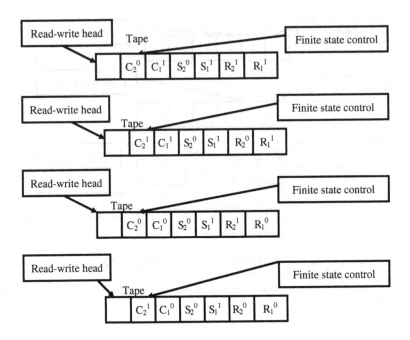

(a) The four **BMDTMs** in the first **BMPDTM**

Fig. 6.96 Schematic representation of the current status of the execution environment to the first **BMPDTM**

6.8 Summary

In this chapter we provided an introduction to how logic operations including **NOT, OR, AND, NOR, NAND, Exclusive-OR (XOR)** and **Exclusive-NOR (XNOR)** on bits were implemented by means of bio-molecular operations. We described **Algorithm 6.1** and **Algorithm 6.2** and their proof to explain how the parallel **NOT** operation of a bit and the parallel **NOT** operation of n bit were constructed by means of biological operations. We also gave one example to show the power to the parallel **NOT** operation of n bit.

We then illustrated **Algorithm 6.3** and **Algorithm 6.4** and their proof to reveal how the parallel **OR** operation of a bit and the parallel **OR** operation of n bit were completed by means of molecular operations. We also gave one example to explain the power to the parallel **OR** operation of n bit. We then introduced **Algorithm 6.5** and **Algorithm 6.6** and their proof to demonstrate how the parallel **AND** operation of a bit and the parallel **AND** operation of n bit were implemented by means of biological operations. We also gave one example to reveal the power to the parallel **AND** operation of n bit.

We then described **Algorithm 6.7** and **Algorithm 6.8** and their proof to show how the parallel **NOR** operation of a bit and the parallel **NOR** operation of n bit were constructed by means of molecular operations. We also gave one example to explain the power to the parallel **NOR** operation of n bit. We then illustrated **Algorithm 6.9** and **Algorithm 6.10** and their proof to reveal how the parallel **NAND** operation of a bit and the parallel **NAND** operation of n bit were completed by means of biological operations. We also gave one example to show the power to the parallel **NAND** operation of n bit.

We then introduced **Algorithm 6.11** and **Algorithm 6.12** and their proof to demonstrate how the parallel **XOR** operation of a bit and the parallel **XOR** operation of n bit were implemented by means of molecular operations. We also gave one example to explain the power to the parallel **XOR** operation of n bit. We then described **Algorithm 6.13** and **Algorithm 6.14** and their proof to show how the parallel **XNOR** operation of a bit and the parallel **XNOR** operation of n bit were constructed by means of biological operations. We also gave one example to reveal the power to the parallel **XNOR** operation of n bit.

6.9 Bibliographical Notes

The textbook by Mano (1979) is a good introduction to logic operations containing **NOT, OR, AND, NOR, NAND, Exclusive-OR (XOR)** and **Exclusive-NOR (XNOR)** on bits. The textbook by Hopcroft et al. (2006) is a good introduction to languages, complexity of computation and automata theory. A good description to Boolean functions discussed in exercises in Sect. 6.10 is Brown and Vranesic (2007), Mano (1979).

6.10 Exercises

6.1 The binary operator $\vee$ defines logical operation **OR**, and the binary operator $\wedge$ defines logical operation **AND**. The truth table to a logical operation, $x \vee (x \wedge y)$, is shown in Table 6.8, where x and y are Boolean variables that are respectively the first input and the second input. Based on Table 6.8, write a bio-molecular program to implement the function of the logical operation, $x \vee (x \wedge y)$.

Table 6.8 The truth table to a logical operation, $x \vee (x \wedge y)$, is shown

The first input (x)	The second input (y)	$x \vee (x \wedge y)$
0	0	0
0	1	0
1	0	1
1	1	1

6.2 The binary operator $\vee$ defines logical operation **OR**, and the binary operator $\wedge$ defines logical operation **AND**. The truth table to a logical operation, $x \wedge (x \vee y)$, is shown in Table 6.9, where x and y are Boolean variables that are subsequently the first input and the second input. Based on Table 6.9, write a bio-molecular program to implement the function of the logical operation, $x \wedge (x \vee y)$.

Table 6.9 The truth table to a logical operation, $x \wedge (x \vee y)$, is shown

The first input (x)	The second input (y)	$x \wedge (x \vee y)$
0	0	0
0	1	0
1	0	1
1	1	1

6.3 The binary operator $\vee$ defines logical operation **OR**, and the binary operator $\wedge$ defines logical operation **AND**. The truth table to a logical operation, $y \vee (y \wedge x)$, is shown in Table 6.10, where x and y are Boolean variables that are subsequently the first input and the second input. Based on Table 6.10, write a bio-molecular program to implement the function of the logical operation, $y \vee (y \wedge x)$.

Table 6.10 The truth table to a logical operation, $y \vee (y \wedge x)$, is shown

The first input (x)	The second input (y)	$y \vee (y \wedge x)$
0	0	0
0	1	1
1	0	0
1	1	1

6.4 The binary operator $\vee$ defines logical operation **OR**, and the binary operator $\wedge$ defines logical operation **AND**. The truth table to a logical operation, $y \wedge (y \vee x)$, is shown in Table 6.11, where x and y are Boolean variables that are respectively the first input and the second input. Based on Table 6.11, write a bio-molecular program to implement the function of the logical operation, $y \wedge (y \vee x)$.

Table 6.11 The truth table to a logical operation, $y \wedge (y \vee x)$, is shown

The first input (x)	The second input (y)	$y \wedge (y \vee x)$
0	0	0
0	1	1
1	0	0
1	1	1

6.5 The unary operator $'$ defines logical operation **NOT**, and the binary operator $\wedge$ defines logical operation **AND**. The truth table to a logical operation, $x' \wedge y$, is shown in Table 6.12, where x and y are Boolean variables that are subsequently the first input and the second input. Based on Table 6.12, write a bio-molecular program to implement the function of the logical operation, $x' \wedge y$.

Table 6.12 The truth table to a logical operation, $x' \wedge y$, is shown	The first input (x)	The second input (y)	$x'' \wedge y$
	0	0	0
	0	1	1
	1	0	0
	1	1	0

6.6 The unary operator $'$ defines logical operation **NOT**, the binary operator $\vee$ defines logical operation **OR**, and the binary operator $\wedge$ defines logical operation **AND**. The truth table to a logical operation, $(x \wedge y) \vee (x' \wedge y')$, is shown in Table 6.13, where x and y are Boolean variables that are respectively the first input and the second input. Based on Table 6.13, write a bio-molecular program to implement the function of the logical operation, $(x \wedge y) \vee (x' \wedge y')$.

Table 6.13 The truth table to a logical operation, $(x \wedge y) \vee (x' \wedge y')$, is shown	The first input (x)	The second input (y)	$(x \wedge y) \vee (x' \wedge y')$
	0	0	1
	0	1	0
	1	0	0
	1	1	1

6.7 The unary operator $'$ defines logical operation **NOT**. The truth table to a logical operation, y, is shown in Table 6.14, where x and y are Boolean variables that are subsequently the first input and the second input. Based on Table 6.14, write a bio-molecular program to implement the function of the logical operation, y'.

Table 6.14 The truth table to a logical operation, y', is shown	The first input (x)	The second input (y)	y'
	0	0	1
	0	1	0
	1	0	1
	1	1	0

6.8 The unary operator $'$ defines logical operation **NOT**. The truth table to a logical operation, x, is shown in Table 6.15, where x and y are Boolean variables that are respectively the first input and the second input. Based on Table 6.15, write a bio-molecular program to implement the function of the logical operation, x'.

Table 6.15 The truth table to a logical operation, x', is shown	The first input (x)	The second input (y)	x'
	0	0	1
	0	1	1
	1	0	0
	1	1	0

6.9 The unary operator $'$ defines logical operation **NOT**, and the binary operator $\vee$ defines logical operation **OR**. The truth table to a logical operation, $x \vee y'$, is shown in Table 6.16, where x and y are Boolean variables that are subsequently the first input and the second input. Based on Table 6.16, write a bio-molecular program to implement the function of the logical operation, $x \vee y'$.

Table 6.16 The truth table to a logical operation, $x \vee y'$, is shown	The first input (x)	The second input (y)	$x \vee y'$
	0	0	1
	0	1	0
	1	0	1
	1	1	1

6.10 The unary operator $'$ defines logical operation **NOT**, and the binary operator $\vee$ defines logical operation **OR**. The truth table to a logical operation, $x' \vee y$, is shown in Table 6.17, where x and y are Boolean variables that are respectively the first input and the second input. Based on Table 6.17, write a bio-molecular program to implement the function of the logical operation, $x' \vee y$

Table 6.17 The truth table to a logical operation, $x' \vee y$, is shown	The first input (x)	The second input (y)	$x' \vee y$
	0	0	1
	0	1	1
	1	0	0
	1	1	1

References

J. Hopcroft, R. Motwani, J. Ullman, *Introduction to Automata Theory, Languages, and Computation* (Addison Wesley, Boston, 2006), ISBN: 81-7808-347-7

M.M. Mano, *Digital Logic and Computer Design* (Prentice-Hall, New York, 1979), ISBN: 0-13-214510-3

S. Brown, Z. Vranesic, *Fundamentals of Digital Logic with Verilog Design* (McGraw-Hill, New York, 2007), ISBN: 978-0077211646

Chapter 7
Introduction to Comparators and Shifters and Increase and Decrease and Two Specific Operations on Bits on Bio-molecular Computer

In previous chapter, we proved how to carry out logic operations on bits in a bio-molecular computer. In this chapter, we will show how to perform comparators, shifters, increase, decrease, two specific operations on bits in a bio-molecular computer. Two specific operations on bits are to find the maximum number of "1" and to find the minimum number of "1". Comparators on bits are, subsequently, ">", "=", "<", "≥", "≤" and "≠". Shifters on bits contain "≪" (a left shifter) and "≫" (a right shifter). The symbol "++" is used to represent to increase for bits, and the symbol "−" is applied to represent to decrease for bits. Those operations on bits are shown in Fig. 7.1. The *bio-molecular parallel deterministic one-tape Turing machine* (abbreviated **BMPDTM**) denoted in Chap. 5 is chosen as our model for the purpose of clearly explaining how those operations on bits are finished.

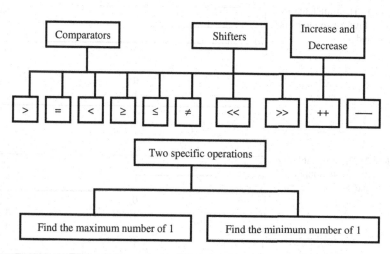

Fig. 7.1 Comparators, shifters, increase, decrease and two specific operations

W.-L. Chang and A. V. Vasilakos, *Molecular Computing*, Studies in Big Data 4, DOI: 10.1007/978-3-319-05122-2_7, © Springer International Publishing Switzerland 2014

7.1 Introduction to Comparators on Bio-molecular Computer

Comparators of a bit for two inputs of a bit, U and V, is used to determine the relationship between U and V by comparing U and V to judge if their values are greater than (">"), equal to ("="), less than ("<"), greater than or equal to ("$\geq$"), less than or equal to ("$\leq$") and unequal to ("$\neq$") each other. The four possible combinations for a comparator ">" of a bit to U and V are: (1) $0 > 0 = 0$, (2) $0 > 1 = 0$, (3) $1 > 0 = 1$ and (4) $1 > 1 = 0$. Similarly, for a comparator "=" of a bit to U and V, the four possible combinations include: (1) $0 = 0 = 1$, (2) $0 = 1 = 0$, (3) $1 = 0 = 0$ and (4) $1 = 1 = 1$. Next, for a comparator "<" of a bit to U and V, the four possible combinations contain: (1) $0 < 0 = 0$, (2) $0 < 1 = 1$, (3) $1 < 0 = 0$ and (4) $1 < 1 = 0$. Similarly, the four possible combinations for a comparator "$\geq$" of a bit to U and V are: (1) $0 \geq 0 = 1$, (2) $0 \geq 1 = 0$, (3) $1 \geq 0 = 1$ and (4) $1 \geq 1 = 1$. Then, for a comparator "$\leq$" of a bit to U and V, the four possible combinations include: (1) $0 \leq 0 = 1$, (2) $0 \leq 1 = 1$, (3) $1 \leq 0 = 0$ and (4) $1 \leq 1 = 1$. Finally, for a comparator "$\neq$" of a bit to U and V, the four possible combinations contain: (1) $0 \neq 0 = 0$, (2) $0 \neq 1 = 1$, (3) $1 \neq 0 = 1$ and (4) $1 \neq 1 = 0$.

Six truth tables are usually used with comparators of a bit to represent all possible combinations of inputs and the corresponding outputs. The six truth tables for ">", "=", "<", "$\geq$", "$\leq$" and "$\neq$" are, subsequently, shown in Tables 7.1, 7.2, 7.3, 7.4, 7.5 and 7.6.

Table 7.1 The truth table for ">" of a bit for two inputs of a bit, U and V

Input		Output
U	V	$U > V$
0	0	0
0	1	0
1	0	1
1	1	0

Table 7.2 The truth table for "=" of a bit for two inputs of a bit, U and V

Input		Output
U	V	$U = V$
0	0	1
0	1	0
1	0	0
1	1	1

Table 7.3 The truth table for "<" of a bit for two inputs of a bit, U and V

Input		Output
U	V	$U < V$
0	0	0
0	1	1
1	0	0
1	1	0

Table 7.4 The truth table for "≥" of a bit for two inputs of a bit, U and V

Input		Output
U	V	$U \geq V$
0	0	1
0	1	0
1	0	1
1	1	1

Table 7.5 The truth table for "≤" of a bit for two inputs of a bit, U and V

Input		Output
U	V	$U \leq V$
0	0	1
0	1	1
1	0	0
1	1	1

Table 7.6 The truth table for "≠" of a bit for two inputs of a bit, U and V

Input		Output
U	V	$U \neq V$
0	0	0
0	1	1
1	0	1
1	1	0

7.1.1 The Construction for the Parallel Comparator of a Bit on Bio-molecular Computer

Suppose that two one-bit binary numbers, U_k and V_k, for $1 \leq k \leq n$ are used to, subsequently, represent the first input and the second input for the comparator of a bit. For the sake of convenience, assume that U_k^1 denotes the fact that the value of U_k is 1 and U_k^0 denotes the fact that the value of U_k is 0. Similarly, suppose that V_k^1 denotes the fact that the value of V_k is 1 and V_k^0 denotes the fact that the value of V_k is 0. The following algorithm is proposed to perform the parallel comparator of a bit.

Algorithm 7.1: One-Bit-ParallelComparator(T_0, T_3, $T_0^>$, $T_0^=$, $T_0^<$, k)

(1) $T_2^{ON} = +(T_0, U_k^1)$ and $T_2^{OFF} = -(T_0, U_k^1)$.

(2) $T_3^{ON} = +(T_3, V_k^1)$ and $T_3^{OFF} = -(T_3, V_k^1)$.

(3) **If** (Detect(T_3^{ON}) $==$ "yes") **then**

      (3a) $T_0^= = \cup(T_0^=, T_2^{ON})$ and $T_0^< = \cup(T_0^<, T_2^{OFF})$.

Else

      (3b) $T_0^= = \cup(T_0^=, T_2^{OFF})$ and $T_0^> = \cup(T_0^>, T_2^{ON})$.

EndIf

(4) $T_3 = \cup(T_3^{ON}, T_3^{OFF})$.

End Algorithm

Lemma 7-1: *The algorithm, **One-Bit-ParallelComparator**(T_0, T_3, $T_0^>$, $T_0^=$, $T_0^<$, k), can be applied to carry out the parallel comparator of a bit.*

Proof The algorithm, **One-Bit-ParallelComparator**(T_0, T_3, $T_0^>$, $T_0^=$, $T_0^<$, k), is implemented by means of the *extract*, *detect* and *merge* operations. Steps (1) and Step (2) use the *extract* operations to form some different test tubes including different inputs. That is, T_2^{ON} contains all of the inputs that have $U_k = 1$, T_2^{OFF} includes all of the inputs that have $U_k = 0$, T_3^{ON} consists of those that have $V_k = 1$, and T_3^{OFF} includes those that have $V_k = 0$.

Step (3) is applied to check whether contains any input for tube T_3^{ON} or not. If any a "yes" is returned from Step (3), then the value of the kth bit for the second input is one. Because the value of the kth bit for the first input in tube T_2^{ON} is one and the value of the kth bit for the first input in tube T_2^{OFF} is zero, from Step (3a), the *merge* operation is applied to pour tube T_2^{ON} into tube $T_0^=$ and also to pour tube T_2^{OFF} into tube $T_0^<$. If any a "no" is returned from Step (3), then the value of the kth bit for the second input is zero. Because the value of the kth bit for the first input in tube T_2^{ON} is one and the value of the kth bit for the first input in tube T_2^{OFF} is zero, from Step (3a), the *merge* operation is applied to pour tube T_2^{ON} into tube $T_0^>$ and also to pour tube T_2^{OFF} into tube $T_0^=$.

Next, on the execution of Step (4), it is employed to pour tubes T_3^{ON} and T_3^{OFF} into tube T_3. This implies that the second input for the comparator of a bit is reserved in tube T_3. Simultaneously, for the comparison of between the kth bit of the first input and the kth bit of the second input, tubes $T_0^>$, $T_0^=$ and $T_0^<$ subsequently contains the comparative result of ">", the comparative result of "=", and the comparative result of "<". Furthermore, truth Tables 7.1, 7.2 and 7.3 are performed. ∎

7.1.2 The Construction for the Parallel Comparator of N Bits on Bio-molecular Computer

The parallel comparator of n bits simultaneously produces the corresponding relationships by means of judging if for 2^n combinations of n bits their values are greater than (">"), equal to ("="), less than ("<"), greater than or equal to ("$\geq$"), less than or equal to ("$\leq$") and unequal to ("$\neq$") each other. The following algorithm is presented to perform the parallel comparator of n bits. Notations in **Algorithm 7.2** are denoted in Sect. 7.1.1.

Algorithm 7.2: N-Bits-ParallelComparator(T_0)

(1) Append-head(T_1, U_1^1).

(2) Append-head(T_2, U_1^0).

(3) $T_0 = \cup(T_1, T_2)$.

(4) **For** $k = 2$ **to** n

    .(4a) Amplify(T_0, T_1, T_2).

    (4b) Append-head(T_1, U_k^1).

    (4c) Append-head(T_2, U_k^0).

    (4d) $T_0 = \cup(T_1, T_2)$.

EndFor

(5) **For** $k = 1$ **to** n

    (5a) Append-head(T_3, V_k).

EndFor

(6) **For** $k = n$ **to** 1

    (6a) **One-Bit-ParallelComparator**(T_0, T_3, $T_0^>$, $T_0^=$, $T_0^<$, k).

    (6b) **If** (Detect($T_0^=$)) == "yes") **then**

        (6c) $T_0 = \cup(T_0, T_0^=)$.

        Else

            (6d) Terminate the execution of the loop.

        EndIf

    EndFor

End Algorithm

Lemma 7-2: *The algorithm*, **N-Bits-ParallelComparator**(T_0), *can be used to carry out the parallel comparator of n bits.*

Proof From Steps (1) through (4d), they are mainly employed to generate solution space of 2^n unsigned integers for the first input (the range of values for them is from 0 to $2^n - 1$). After they are finished, tube T_0 contains 2^n combinations of n bits. Next, Step (5) is the second loop and each execution of Step (5a) is applied to append the value "0" or "1" for V_k (the kth bit of the second input) onto the head of the bit pattern, $V_{k-1}, \ldots, V_1$, in tube T_3. Step (6) is the third loop and is mainly applied to carry out the parallel comparator of n bits. Each execution of Step (6a) calls **One-Bit-ParallelComparator**(T_0, T_3, $T_0^>$, $T_0^=$, $T_0^<$, k) in Sect. 7.1.1 to perform the comparison for the kth bit of 2^n input pairs ($U_n, \ldots, U_1$, $V_n, \ldots, V_1$). On each execution of Step (6b), it uses the *detect* operation to test if there is any an input in tube $T_0^=$. If a "yes" is returned, then from Step (6c) in tube $T_0^=$ those inputs ($U_n, \ldots, U_1$) which comparative result of the kth bit is equal to ("=") are poured into tube T_0. Otherwise, from Step (6d), the execution of the loop is terminated. Repeat execution of Step (6a) until the comparison for the nth bit of 2^n input pairs is finished. Tube $T_0^>$ includes those inputs $(U_n, \ldots, U_1)$ which comparative result is greater than ("$>$"), tube $T_0^=$ contains those inputs $(U_n, \ldots, U_1)$ which comparative result is equal to ("$=$"), and tube $T_0^<$ consists of those inputs $(U_n, \ldots, U_1)$ which comparative result is less than ("$<$").   ∎

7.1.3 The Power for the Parallel Comparator of N Bits on Bio-molecular Computer

Consider that four values for an unsigned integer of two bits are, subsequently, $00(0_{10})$ $(U_2^0 U_1^0)$, $01(1_{10})$ $(U_2^0 U_1^1)$, $10(2_{10})$ $(U_2^1 U_1^0)$ and $11(3_{10})$ $(U_2^1 U_1^1)$. For any given value $01(1_{10})$ $(V_2^0 V_1^1)$, we want to simultaneously find various relationships among $01(1_{10})$ $(V_2^0 V_1^1)$ and those four values each other. **Algorithm 7.2**, **N-Bits-ParallelComparator**(T_0), can be employed to perform the task. Tube T_0 is an empty tube and is regarded as an input tube of **Algorithm 7.2**. From Definition 5−2, the input tube T_0 is regarded as the execution environment of the first **BMPDTM**. Similarly, tubes T_1, T_2 and T_3 used in **Algorithm 7.2** also are regarded, subsequently, as the execution environment of the second **BMPDTM**, the execution environment of the third **BMPDTM** and the execution environment of the fourth **BMPDTM**.

Steps (1) through (4d) in **Algorithm 7.2** are used to generate a **BMPDTM** with four bio-molecular deterministic one-tape Turing machines. After the execution for Step (1) and Step (2) of **Algorithm 7.2** is finished, tube $T_1 = \{U_1^1\}$ and tube $T_2 = \{U_1^0\}$. This is to say that a **BMDTM** in the second **BMPDTM** and in the third **BMPDTM** is constructed. Figure 7.2 is used to reveal the current status of the execution environment to the second **BMPDTM** and the third **BMPDTM**. From Fig. 7.2, the content of the first tape square for the tape in the first **BMDTM** in the second **BMPDTM** is written by its corresponding read-write head and is 1 $(U_1 = 1)$, and the content of the first tape square for the tape in the first **BMDTM** in the third **BMPDTM** is written by its corresponding read-write head and is 0 $(U_1 = 0)$. For the first **BMDTM** in the second **BMPDTM**, the position of the read-write head is moved to the *left new* tape square, and the state of the finite state control is changed as "$U_1 = 1$". Similarly, for the first **BMDTM** in the third **BMPDTM**, the position of the read-write head is moved to the *left new* tape square, and the state of the finite state control is changed as "$U_1 = 0$".

Next, after the execution for Step (3) of **Algorithm 7.2** is performed, tube $T_0 = \{U_1^1, U_1^0\}$, tube $T_1 = \varnothing$ and tube $T_2 = \varnothing$. This is to say that the execution environment for the first bio-molecular deterministic one-tape Turing machine in the second **BMPDTM** and the first bio-molecular deterministic one-tape Turing machine in the third **BMPDTM** becomes the first **BMPDTM**. The position of the corresponding read-write head and the state of the corresponding finite state control are both reserved. Figure 7.3 is applied to show the current status of the execution environment to the first **BMPDTM**. From Fig. 7.3, it is pointed out that the contents to the two tapes in the execution environment of the first **BMPDTM** are not changed.

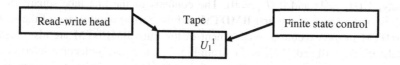

(a) The first **BMDTM** in the second **BMPDTM**.

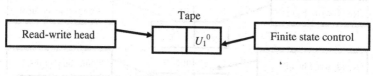

(b) The first **BMDTM** in the third **BMPDTM**.

Fig. 7.2 Schematic representation of the current status of the execution environment to the second **BMPDTM** and the third **BMPDTM**

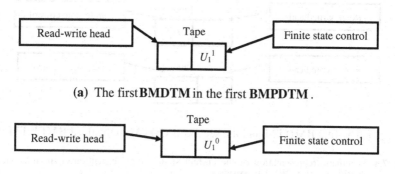

(a) The first **BMDTM** in the first **BMPDTM**.

(b) The second **BMDTM** in the first **BMPDTM**.

Fig. 7.3 Schematic representation of the current status of the execution environment to the first **BMPDTM**

Step (4) is the first loop in **Algorithm 7.2**, since the number of bits for representing those four values is two, the upper bound (n) is two. Therefore, after the first execution of Step (4a) is implemented, tube $T_0 = \varnothing$, tube $T_1 = \{U_1{}^1, U_1{}^0\}$ and tube $T_2 = \{U_1{}^1, U_1{}^0\}$. This indicates that the first **BMDTM** and the second **BMDTM** in the execution environment of the first **BMPDTM** are both copied into the second **BMPDTM** and the third **BMPDTM**. Figure 7.4 is applied to explain

the current status of the execution environment to the second **BMPDTM** and the
third **BMPDTM**. From Fig. 7.4, the contents of the first tape square for the corresponding tape of the first **BMDTM** and the corresponding tape of the second
BMDTM in the execution environment of the second **BMPDTM** are, respectively, 1 ($U_1 = 1$) and 0 ($U_1 = 0$). The contents of the first tape square for the
corresponding tape of the first **BMDTM** and the corresponding tape of the second
BMDTM in the execution environment of the third **BMPDTM** are also, respectively, 0 ($U_1 = 0$) and 1 ($U_1 = 1$). From Fig. 7.4, four bio-molecular deterministic
one-tape Turing machines are generated. For the four bio-molecular deterministic

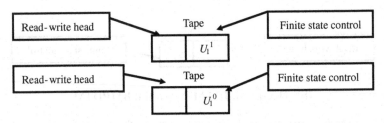

(a) The first **BMDTM** and the second **BMDTM** in the second **BMPDTM**.

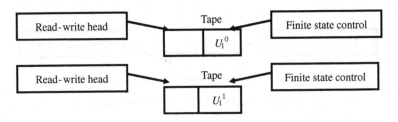

(b) The first **BMDTM** and the second **BMDTM** in the third **BMPDTM**.

Fig. 7.4 Schematic representation of the current status of the execution environment to the
second **BMPDTM** and the third **BMPDTM**

one-tape Turing machines, the position of the corresponding read-write head and
the state of the corresponding finite state control are reserved.

Next, after the first execution for Step (4b) and Step (4c) of **Algorithm 7.2** is
finished, tube $T_1 = \{U_2{}^1 U_1{}^1, U_2{}^1 U_1{}^0\}$ and tube $T_2 = \{U_2{}^0 U_1{}^1, U_2{}^0 U_1{}^0\}$. This is
to say that the content of the *second* tape square for the tape in the first **BMDTM**
in the second **BMPDTM** is written by its corresponding read-write head and is 1
($U_2 = 1$), and the content of the *second* tape square for the tape in the second
BMDTM in the second **BMPDTM** is written by its corresponding read-write head
and is also 1 ($U_2 = 1$). Simultaneously, the position of the corresponding read-
write head is both moved to the *left new* tape square and the state of the

corresponding finite state control is both changed as "$U_2 = 1$". Similarly, the content of the *second* tape square for the tape in the first **BMDTM** in the third **BMPDTM** is written by its corresponding read-write head and is 0 ($U_2 = 0$), and the content of the *second* tape square for the tape in the second **BMDTM** in the third **BMPDTM** is written by its corresponding read-write head and is also 0 ($U_2 = 0$). Simultaneously, the position of the corresponding read-write head is both moved to the *left new* tape square and the state of the corresponding finite state control is both changed as "$U_2 = 0$". Figure 7.5 is used to reveal the current status of the execution environment to the second **BMPDTM** and the third **BMPDTM**.

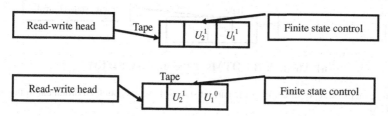

(a) The first **BMDTM** and the second **BMDTM** in the second **BMPDTM**.

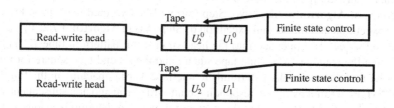

(b) The first **BMDTM** and the second **BMDTM** in the third **BMPDTM**.

Fig. 7.5 Schematic representation of the current status of the execution environment to the second **BMPDTM** and the third **BMPDTM**

Next, after the first execution for Step (4d) of **Algorithm 7.2** is implemented, tube $T_0 = \{U_2{}^1 U_1{}^1, U_2{}^1 U_1{}^0, U_2{}^0 U_1{}^1, U_2{}^0 U_1{}^0\}$, tube $T_1 = \emptyset$ and tube $T_2 = \emptyset$. This indicates that the execution environment for those bio-molecular deterministic one-tape Turing machines in the second **BMPDTM** and in the third **BMPDTM** becomes the first **BMPDTM**. Figure 7.6 is used to illustrate the current status of the execution environment to the first **BMPDTM**. From Fig. 7.6, the contents to the four tapes in the execution environment of the first **BMPDTM** are not changed, and the position of the corresponding read-write head and the state of the corresponding finite state control are reserved.

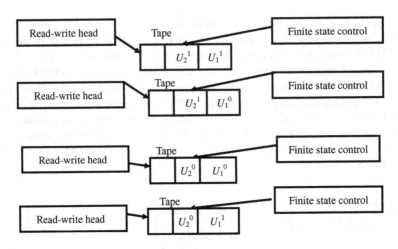

(a) The four **BMDTMs** in the first **BMPDTM**.

Fig. 7.6 Schematic representation of the current status of the execution environment to the first **BMPDTM**

Then, because Step (5) of **Algorithm 7.2** is the second loop and the upper bound for Step (5) of **Algorithm 7.2** is two, Step (5a) will be executed two times. From the first execution and the second execution of Step (5a) in **Algorithm 7.2**, V_1^1 and V_2^0 are, subsequently, appended into the head of each bit pattern in tube T_3. This implies that the contents of the first tape square and the second tape square for the tape of the first bio-molecular deterministic one-tape Turing machine in the fourth **BMPDTM** are written by the corresponding read-write head and are, subsequently, V_1^1 and V_2^0. Simultaneously, for the first bio-molecular deterministic one-tape Turing machine in the fourth **BMPDTM**, the position of the corresponding read-write head is moved to the *left new* tape square, the state of the corresponding finite state control is changed as "$V_2 = 0$" and tube $T_3 = \{V_2^0 \, V_1^1\}$. Figure 7.7 is employed to show the current status of the execution environment to the fourth **BMPDTM**.

Step (6) in **Algorithm 7.2** is a loop and is employed to perform the parallel comparator of n bits. When the first execution of Step (6a) is finished, it calls **Algorithm 7.1** that is applied to carry out the parallel comparator of one bit, **One-Bit-ParallelComparator** $(T_0, T_3, T_0^>, T_0^=, T_0^<, k)$, in Sect. 7.1.1. The first parameter, tube T_0, is current the execution environment of the first **BMPDTM** and includes four bio-molecular deterministic one-tape Turing machines (Fig. 7.6). It is regarded as an input tube of **Algorithm 7.1**. The second parameter, tube T_3, is current the execution environment of the fourth **BMPDTM** and includes one bio-molecular deterministic one-tape Turing machine (Fig. 7.7). It is regarded as an input tube of **Algorithm 7.1**. Tubes $T_0^>$, $T_0^=$ and $T_0^<$ are empty tubes and are regarded as input tubes of **Algorithm 7.1**. The value for the six parameter, k, is two and is also regarded as an input value of **Algorithm 7.1**.

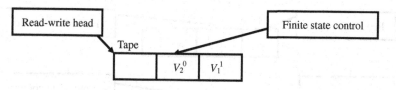

(a) The first **BMDTM** in the fourth **BMPDTM**.

Fig. 7.7 Schematic representation of the current status of the execution environment to the fourth **BMPDTM**

When **Algorithm 7.1** is first invoked, nine tubes are all regarded as independent environments of nine bio-molecular parallel deterministic one-tape Turing machines. Tube T_0 is regarded as the first **BMPDTM** and tubes T_3, T_2^{ON}, T_2^{OFF}, T_3^{ON}, T_3^{OFF}, $T_0^>$, $T_2^=$ and $T_2^<$ are, subsequently, the fourth **BMPDTM**, the fifth **BMPDTM**, the sixth **BMPDTM**, the seventh **BMPDTM**, the eighth **BMPDTM**, the nineth **BMPDTM**, the tenth **BMPDTM** and the eleventh **BMPDTM**. After the *first* execution of Step (1) in **Algorithm 7.1** is implemented, tube $T_0 = \varnothing$, tube $T_2^{ON} = \{U_2^1 U_1^1, U_2^0 U_1^1\}$ and tube $T_2^{OFF} = \{U_2^1 U_1^0, U_2^0 U_1^0\}$. This is to say that the new execution environments for two bio-molecular deterministic one-tape Turing machines with the content of tape square, "U_1^1", and other two bio-molecular deterministic one-tape Turing machines with the content of tape square, "U_1^0" are, respectively, the fifth **BMPDTM** and the sixth **BMPDTM**. The position of the corresponding read-write head and the state of the corresponding finite state control are reserved. Figures 7.8 and 7.9 are applied to reveal the result.

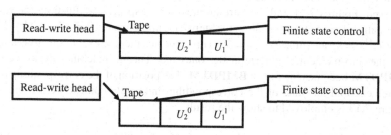

(a) The two **BMDTMs** in the fifth **BMPDTM**.

Fig. 7.8 Schematic representation of the current status of the execution environment to the fifth **BMPDTM**

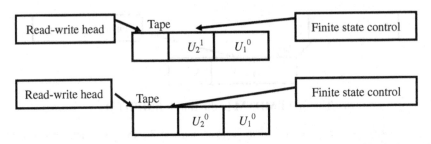

(a) The two **BMDTM**s in the sixth **BMPDTM**.

Fig. 7.9 Schematic representation of the current status of the execution environment to the sixth **BMPDTM**

Then, after the *first* execution of Step (2) in **Algorithm 7.1** is performed, tube $T_3 = \varnothing$, tube $T_3^{ON} = \varnothing$, and tube $T_3^{OFF} = \{V_2^0\ V_1^1\}$. This implies that the new execution environment for the bio-molecular deterministic one-tape Turing machine with the contents of tape square, "V_2^0", is the eighth **BMPDTM**. The position of the corresponding read-write head and the state of the corresponding finite state control are reserved. Figure 7.10 is employed to reveal the result.

Next, after the first execution for Step (3) is finished, the returned result from Step (3) is "no". Therefore, after the first execution for Step (3b) is implemented, tube $T_2^{ON} = \varnothing$, tube $T_2^{OFF} = \varnothing$, tube $T_0^= = \{U_2^0\ U_1^1,\ U_2^0\ U_1^0\}$ and tube $T_0^> = \{U_2^1\ U_1^1,\ U_2^1\ U_1^0\}$. This is to say that the execution environments for the four bio-molecular deterministic one-tape Turing machines, respectively, become the tenth **BMPDTM** and the nineth **BMPDTM**. The position of the corresponding read-write head and the state of the corresponding finite state control are all reserved. Figures 7.11 and 7.12 are employed to respectively illustrate the result.

Finally, after the first execution for Step (4) is implemented, tube $T_3 = \{V_2^0\ V_1^1\}$, tube $T_3^{ON} = \varnothing$ and tube $T_3^{OFF} = \varnothing$. This indicates that the execution environment for the bio-molecular deterministic one-tape Turing machine in the eighth **BMPDTM** becomes the fourth **BMPDTM**. The position of the corresponding read-write head and the state of the corresponding finite state control are all reserved. Figure 7.13 is employed to show the result.

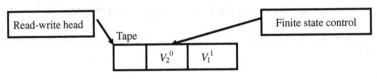

(a) The **BMDTM** in the eighth **BMPDTM**.

Fig. 7.10 Schematic representation of the current status of the execution environment to the eighth **BMPDTM**

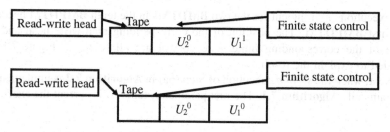

(a) The two **BMDTMs** in the tenth **BMPDTM**.

Fig. 7.11 Schematic representation of the current status of the execution environment to the tenth **BMPDTM**

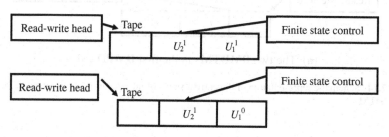

(a) The two **BMDTMs** in the ninth **BMPDTM**.

Fig. 7.12 Schematic representation of the current status of the execution environment to the nineth **BMPDTM**

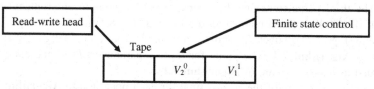

(a) The first **BMDTM** in the fourth **BMPDTM**.

Fig. 7.13 Schematic representation of the current status of the execution environment to the fourth **BMPDTM**

After the execution of the *first* time for each operation in **Algorithm 7.1** is performed, the parallel comparator to the first bit of those inputs is also performed. Next, after the first execution of Step (6b) is performed, because tube $T_0^= = \{U_2^0 \, U_1^1, U_2^0 \, U_1^0\}$, a "yes" is returned. Therefore, then, after the first execution of Step (6c) is performed, tube $T_0^= = \varnothing$ and $T_0 = \{U_2^0 \, U_1^1, U_2^0 \, U_1^0\}$. This is to say that the execution environment for the two **BMDTMs** in the tenth **BMPDTM** becomes the first **BMPDTM**. The position of the corresponding read-write head and the state of the corresponding finite state control are all reserved. Figure 7.14 is applied to explain the result.

Next, when the *second* execution of Step (6a) in **Algorithm 7.2** is implemented, it again calls **Algorithm 7.1**. The first parameter, tube T_0, is current the execution

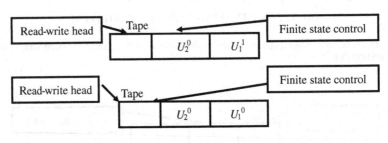

(a) The two **BMDTMs** in the first **BMPDTM**.

Fig. 7.14 Schematic representation of the current status of the execution environment to the first **BMPDTM**

environment of the first **BMPDTM** and consists of two bio-molecular deterministic one-tape Turing machines (Fig. 7.14). It is regarded as an input tube of **Algorithm 7.1**. The second parameter, tube T_3, is current the execution environment of the fourth **BMPDTM** and includes one bio-molecular deterministic one-tape Turing machine (Fig. 7.13). It is also regarded as an input tube of **Algorithm 7.1**. The third parameter, tube $T_0^>$, is current the execution environment of the nineth **BMPDTM** and contains two bio-molecular deterministic one-tape Turing machine (Fig. 7.12). Tubes $T_0^=$ and $T_0^<$ are empty tubes and are regarded as input tubes of **Algorithm 7.1**. The value for the six parameter, k, is one and is also regarded as an input value of **Algorithm 7.1**.

After the execution of the *second* time for each operation in **Algorithm 7.1** is finished, tube $T_0 = \varnothing$, tube $T_3 = \{V_2^0\,V_1^1\}$, tube $T_0^> = \{U_2^1\,U_1^1,\,U_2^1\,U_1^0\}$, tube $T_0^= = \{U_2^0\,U_1^1\}$ and tube $T_0^< = \{U_2^0\,U_1^0\}$ and other tubes become all empty tubes. Figure 7.15 is used to illustrate the result and **Algorithm 7.2** is terminated.

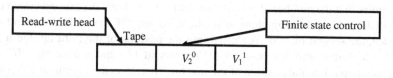

(a) The first **BMDTM** in the fourth **BMPDTM**.

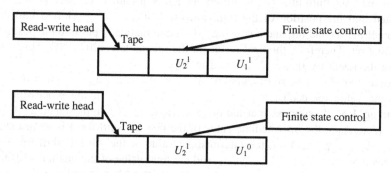

(b) The two **BMDTM**s in the ninth **BMPDTM**.

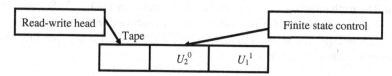

(c) The first **BMDTM** in the tenth **BMPDTM**.

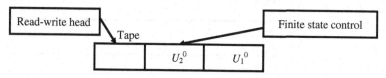

(d) The first **BMDTM** in the eleventh **BMPDTM**.

Fig. 7.15 Schematic representation of the current status of the execution environment to the fourth **BMPDTM**, the nineth **BMPDTM**, the tenth **BMPDTM** and the eleventh **BMPDTM**

7.2 Introduction to Left Shifters on Bio-molecular Computer

The left shifter is applied to compute $A \times 2^B$, where A and B are unsigned integers of n bits and B is used to represent the number of bits for left shift. A symbol "$\ll$" is used to represent the operation of left shift, and the expression $A \times 2^B$ can be rewritten as another expression: $A \ll B$. Consider how to perform the computational task of 001×2^2. The expression 001×2^2 can be rewritten as $001 \ll 2$. This implies that the number of bits for left shift is two. When the first execution of left shift for 001×2^2 is implemented, the third bit, "0", is shifted out and is discarded, the second bit, "0" is shifted into the position of the third bit, the first bit, "1" is shifted into the position of the second bit, and a new bit, "0" is automatically put into the position of the first bit. Therefore, the intermediate result for the first left shift is 010. Next, when the second execution of left shift is implemented, the third bit, "0", is shifted out and is discarded, the second bit, "1" is shifted into the position of the third bit, the first bit, "0" is shifted into the position of the second bit, and a new bit, "0" is automatically put into the position of the first bit. Therefore, the final result for the second left shift is 100. This is to say that the result for 001×2^2 is equal to 100.

Assume that A can be represented as $A_{d, n}, A_{d, n-1}, \ldots, A_{d, 1}$ for $0 \le d \le B$, where the value for each $A_{d, k}$ for $1 \le k \le n$ is 1 or 0. In the processing of performing $A \times 2^B$, suppose that the *original* value for A is represented as $A_{0, n}, A_{0, n-1}, \ldots, A_{0, 1}$, the intermediate value of the first left shift for A is represented as $A_{1, n}, A_{1, n-1}, \ldots, A_{1, 1}$, the intermediate value of the dth left shift for A is represented as $A_{d, n}, A_{d, n-1}, \ldots, A_{d, 1}$, and the final value of the *last* left shift for A is represented as $A_{B, n}, A_{B, n-1}, \ldots, A_{B, 1}$. One truth table is usually applied with a left shifter to represent all possible combinations of inputs and the corresponding outputs. The truth table for "$\ll$" (a left shifter) is shown in Table 7.7. Because the bit $A_{d-1, n}$ is shifted out and is discarded, it is not contained in an input column in Table 7.7. Similarly, the bit $A_{d, 1}$ is filled with a bit "0", so it is also not included in an output column in Table 7.7.

Table 7.7 The truth table for the dth left shift to $A \times 2^B$	Input	Output
	$A_{d-1, k}$	$A_{d, k+1}$
	0	0
	1	1

7.2.1 The Construction for the Parallel Left Shifter on Biomolecular Computer

For the sake of convenience, assume that $A_{d,k}{}^1$ for $0 \le d \le B$ and $1 \le k \le n$ denotes the fact that the value of $A_{d, k}$ is 1 and $A_{d, k}{}^0$ for $0 \le d \le B$ and $1 \le k \le n$ denotes the fact that the value of $A_{d, k}$ is 0. The following algorithm is offered to carry out the parallel left shifter. Some notations in **Algorithm 7.3** are denoted in Sects. 7.2 and 7.2.1.

Algorithm 7.3: Parallel-Left-Shifter(T_0)

(1) Append-head($T_1, A_{0,1}^1$).

(2) Append-head($T_2, A_{0,1}^0$).

(3) $T_0 = \cup(T_1, T_2)$.

(4) **For** $k = 2$ **to** n

    (4a) Amplify(T_0, T_1, T_2).

    (4b) Append-head($T_1, A_{0,k}^1$).

    (4c) Append-head($T_2, A_{0,k}^0$).

    (4d) $T_0 = \cup(T_1, T_2)$.

EndFor

(5) **For** $d = 1$ **to** B

    (5a) Append-head($T_0, A_{d,1}^0$).

    (6) **For** $k = 1$ **to** $n - 1$

        (6a) $T_3 = +(T_0, A_{d-1,k}^1)$ and $T_4 = -(T_0, A_{d-1,k}^1)$.

        (6b) **If** (Detect(T_3) = = "yes") **Then**

            (6c) Append-head($T_3, A_{d,k+1}^1$).

    EndIf

        (6d) **If** (Detect(T_4) = = "yes") **Then**

            (6e) Append-head($T_4, A_{d,k+1}^0$).

    EndIf

        (6f) $T_0 = \cup(T_3, T_4)$.

    EndFor

EndFor

End Algorithm

Lemma 7-3: *The algorithm, **Parallel-Left-Shifter(T_0)**, can be used to perform the parallel left shifter.*

Proof The algorithm, **Parallel-Left-Shifter(T_0)**, is implemented by means of the *extract, detect, append-head* and *merge* operations. From Steps (1) through (4d), they are mainly used to construct solution space of 2^n unsigned integers of n bits for the only input (the range of values for them is from 0 to $2^n - 1$). After they are performed, tube T_0 contains 2^n combinations of n bits.

Step (5) is a nested loop and is employed to perform the parallel left shift of B times for 2^n combinations of n bits in tube T_0. On each execution of Step (5a), it uses the *append-head* operation to put the value "0" into the position of the first bit to each left shift. Step (6) is the inner loop and is mainly applied to perform the left shift of one time. Each execution of Step (6a) applies the *extract* operation to form two different tubes including different inputs. That is, T_3 contains all of the inputs that have $A_{d-1,k} = 1$ and T_4 includes all of the inputs that have $A_{d-1,k} = 0$. Next, on each execution of Step (6b) and Step (6d), the *detect* operations are applied to test if contains any input for tubes T_3 and T_4. If a "yes" is returned from Step (6b), then each execution of Step (6c) employs the *append-head* operation to put the value "1" into the position of the $(k + 1)$th bit ($A_{d,k+1} = 1$) in tube T_3 for the left shift of the dth time. Similarly, if a "yes" is returned from Step (6d), then each execution of Step (6e) uses the *append-head* operation to put the value "0" into the position of the $(k + 1)$th bit ($A_{d, k + 1} = 0$) in tube T_4 for the left shift of the dth time. Then, on each execution of Step (6f), it uses the *merge* operation to pour tubes T_3 and T_4 into tube T_0. Repeat execution of each operation is the nested loop until left shift of the *last* time is performed. Each input in tube T_0 performs its value by 2^B.   ∎

7.2.2 The Power for the Parallel Left Shifters of N Bits on Bio-molecular Computer

Consider that four values for an unsigned integer of two bits are, subsequently, $00(0_{10})$ $(A_{0, 2}{}^0 A_{0, 1}{}^0)$, $01(1_{10})$ $(A_{0, 2}{}^0 A_{0, 1}{}^1)$, $10(2_{10})$ $(A_{0, 2}{}^1 A_{0, 1}{}^0)$ and $11(3_{10})$ $(A_{0, 2}{}^1 A_{0, 1}{}^1)$. For any given value 2^1, we want to simultaneously perform $(A_{0, 2}{}^0 A_{0, 1}{}^0) \times 2^1$, $(A_{0, 2}{}^0 A_{0, 1}{}^1) \times 2^1$, $(A_{0, 2}{}^1 A_{0, 1}{}^0) \times 2^1$, and $(A_{0, 2}{}^1 A_{0, 1}{}^1) \times 2^1$. **Algorithm 7.3**, **Parallel-Left-Shifter**(T_0), can be applied to carry out the computational task. Tube T_0 is an empty tube and is regarded as an input tube of **Algorithm 7.3**. According to Definition 5–2, the input tube T_0 is regarded as the execution environment of the first **BMPDTM**. Similarly, tubes T_1, T_2, T_3 and T_4 used in **Algorithm 7.3** also are regarded, subsequently, as the execution environment of the second **BMPDTM**, the execution environment of the third **BMPDTM**, the execution environment of the fourth **BMPDTM**, and the execution environment of the fifth **BMPDTM**.

Steps (1) through (4d) in **Algorithm 7.3** are applied to construct a **BMPDTM** with four bio-molecular deterministic one-tape Turing machines. After the execution for Step (1) and Step (2) of **Algorithm 7.3** is performed, tube $T_1 = \{A_{0, 1}{}^1\}$ and tube $T_2 = \{A_{0, 1}{}^0\}$. This implies that a **BMDTM** in the second **BMPDTM** and in the third **BMPDTM** is constructed. Figure 7.16 is employed to illustrate the current status of the execution environment to the second **BMPDTM** and the third **BMPDTM**. From Fig. 7.16, the content of the first tape square for the tape in the first **BMDTM** in the second **BMPDTM** is written by its corresponding read-write head and is 1 ($A_{0, 1} = 1$), and the content of the first tape square for the tape in the first **BMDTM** in the third **BMPDTM** is written by its corresponding read-write head and is 0 ($A_{0, 1} = 0$). For the first **BMDTM** in the second **BMPDTM**, the position of the read-write head is moved to the *left new* tape square, and the state of the finite state control is changed as "$A_{0, 1} = 1$". Similarly, for the first **BMDTM** in the third **BMPDTM**, the position of the read-write head is moved to the *left new* tape square, and the state of the finite state control is changed as "$A_{0, 1} = 0$".

Next, after the execution for Step (3) of **Algorithm 7.3** is implemented, tube $T_0 = \{A_{0, 1}{}^1, A_{0, 1}{}^0\}$, tube $T_1 = \varnothing$ and tube $T_2 = \varnothing$. This indicates that the execution environment for the first bio-molecular deterministic one-tape Turing machine in the second **BMPDTM** and the first bio-molecular deterministic one-tape Turing machine in the third **BMPDTM** becomes the first **BMPDTM**. The position of the corresponding read-write head and the state of the corresponding finite state control are both reserved. Figure 7.17 is used to reveal the current status of the execution environment to the first **BMPDTM**. From Fig. 7.17, it is indicated that the contents to the two tapes in the execution environment of the first **BMPDTM** are not changed.

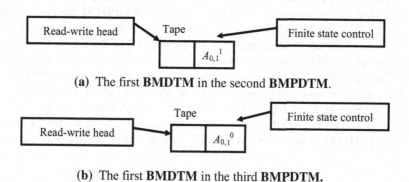

(a) The first **BMDTM** in the second **BMPDTM**.

(b) The first **BMDTM** in the third **BMPDTM**.

Fig. 7.16 Schematic representation of the current status of the execution environment to the second**BMPDTM** and the third **BMPDTM**

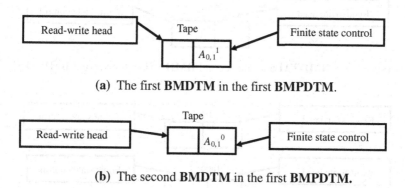

(a) The first **BMDTM** in the first **BMPDTM**.

(b) The second **BMDTM** in the first **BMPDTM**.

Fig. 7.17 Schematic representation of the current status of the execution environment to the first **BMPDTM**

Step (4) is the first loop in **Algorithm 7.3**, because the number of bits for representing those four values is two, the upper bound (n) is two. Thus, after the first execution of Step (4a) is finished, tube $T_0 = \varnothing$, tube $T_1 = \{A_{0,1}{}^1, A_{0,1}{}^0\}$ and tube $T_2 = \{A_{0,1}{}^1, A_{0,1}{}^0\}$. This is to say that the first **BMDTM** and the second

BMDTM in the execution environment of the first **BMPDTM** are both copied into the second **BMPDTM** and the third **BMPDTM**. Figure 7.18 is used to show the current status of the execution environment to the second **BMPDTM** and the third **BMPDTM**. From Fig. 7.18, the contents of the first tape square for the corresponding tape of the first **BMDTM** and the corresponding tape of the second **BMDTM** in the execution environment of the second **BMPDTM** are, respectively, 1 ($A_{0, 1} = 1$) and 0 ($A_{0, 1} = 0$). The contents of the first tape square for the corresponding tape of the first **BMDTM** and the corresponding tape of the second **BMDTM** in the execution environment of the third **BMPDTM** are also, respectively, 0 ($A_{0, 1} = 0$) and 1 ($A_{0, 1} = 1$). From Fig. 7.18, four bio-molecular deterministic one-tape Turing machines are produced. For the four bio-molecular deterministic one-tape Turing machines, the position of the corresponding read-write head and the state of the corresponding finite state control are reserved.

Next, after the first execution for Step (4b) and Step (4c) of **Algorithm 7.3** is implemented, tube $T_1 = \{A_{0, 2}{}^1 A_{0, 1}{}^1, A_{0, 2}{}^1 A_{0, 1}{}^0\}$ and tube $T_2 = \{A_{0, 2}{}^0 A_{0, 1}{}^1, A_{0, 2}{}^0 A_{0, 1}{}^0\}$. This implies that the content of the *second* tape square for the tape in the first **BMDTM** in the second **BMPDTM** is written by its corresponding read-

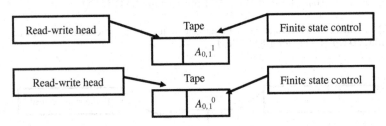

(a) The first **BMDTM** and the second **BMDTM** in the second **BMPDTM**.

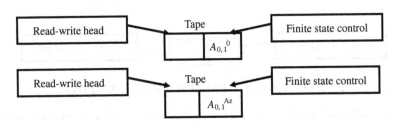

(b) The first **BMDTM** and the second **BMDTM** in the third **BMPDTM**.

Fig. 7.18 Schematic representation of the current status of the execution environment to the second **BMPDTM** and the third **BMPDTM**

write head and is 1 ($A_{0, 2} = 1$), and the content of the *second* tape square for the tape in the second **BMDTM** in the second **BMPDTM** is written by its corresponding read-write head and is also 1 ($A_{0, 2} = 1$). Simultaneously, the position of the corresponding read-write head is both moved to the *left new* tape square and the state of the corresponding finite state control is both changed as "$A_{0, 2} = 1$". Similarly, the content of the *second* tape square for the tape in the first **BMDTM** in the third **BMPDTM** is written by its corresponding read-write head and is 0 ($A_{0, 2} = 0$), and the content of the *second* tape square for the tape in the second **BMDTM** in the third **BMPDTM** is written by its corresponding read-write head and is also 0 ($A_{0, 2} = 0$). Simultaneously, the position of the corresponding read-write head is both moved to the *left new* tape square and the state of the corresponding finite state control is both changed as "$A_{0, 2} = 0$". Figure 7.19 is employed to explain the current status of the execution environment to the second **BMPDTM** and the third **BMPDTM**.

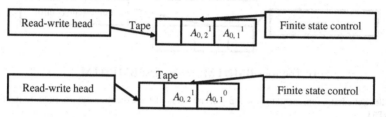

(a) The first **BMDTM** and the second **BMDTM** in the second **BMPDTM**.

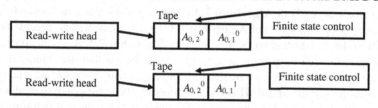

(b) The first **BMDTM** and the second **BMDTM** in the third **BMPDTM**.

Fig. 7.19 Schematic representation of the current status of the execution environment to the second **BMPDTM** and the third **BMPDTM**

Next, after the first execution for Step (4d) of **Algorithm 7.3** is performed, tube $T_0 = \{A_{0, 2}{}^1 A_{0, 1}{}^1, A_{0, 2}{}^1 A_{0, 1}{}^0, A_{0, 2}{}^0 A_{0, 1}{}^1, A_{0, 2}{}^0 A_{0, 1}{}^0\}$, tube $T_1 = \varnothing$ and tube $T_2 = \varnothing$. This is to say that the execution environment for those bio-molecular deterministic one-tape Turing machines in the second **BMPDTM** and in the third **BMPDTM** becomes the first **BMPDTM**. Figure 7.20 is applied to show the current status of the execution environment to the first **BMPDTM**. From Fig. 7.20, the contents to the four tapes in the execution environment of the first **BMPDTM** are not changed, and the position of the corresponding read-write head and the state of the corresponding finite state control are reserved.

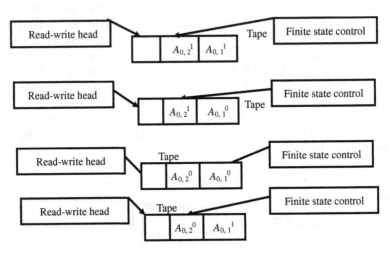

(a) The four **BMDTMs** in the first **BMPDTM**.

Fig. 7.20 Schematic representation of the current status of the execution environment to the first **BMPDTM**

Then, since Step (5) of **Algorithm 7.3** is the nested loop and the upper bound for the outer loop in Step (5) of **Algorithm 7.3** is 1, Step (5a) will be executed one time. From the first execution of Step (5a) in **Algorithm 7.3**, $A_{1, 1}{}^{0}$ is appended into the head of each bit pattern in tube T_0. This indicates that the content of the third tape square for each tape of the four bio-molecular deterministic one-tape Turing machines in the first **BMPDTM** is written by the corresponding read-write head and is 0 ($A_{1, 1} = 0$). Simultaneously, for the four bio-molecular deterministic one-tape Turing machines in the first **BMPDTM**, the position of the corresponding read-write head is moved to the *left new* tape square, the state of the corresponding finite state control is changed as "$A_{1, 1} = 0$" and tube $T_0 = \{A_{1, 1}{}^{0} A_{0, 2}{}^{1} A_{0, 1}{}^{1}, A_{1, 1}{}^{0} A_{0, 2}{}^{1} A_{0, 1}{}^{0}, A_{1, 1}{}^{0} A_{0, 2}{}^{0} A_{0, 1}{}^{1}, A_{1, 1}{}^{0} A_{0, 2}{}^{0} A_{0, 1}{}^{0}\}$. Figure 7.21 is used to reveal the current status of the execution environment to the first **BMPDTM**.

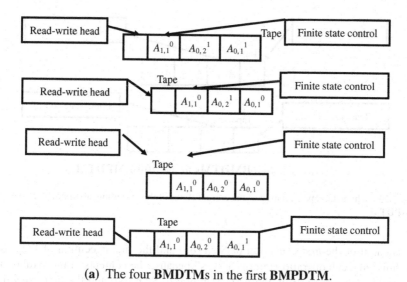

(a) The four **BMDTMs** in the first **BMPDTM**.

Fig. 7.21 Schematic representation of the current status of the execution environment to the first **BMPDTM**

Step (6) in **Algorithm 7.3** is the inner loop, its upper bound is one, and is applied to carry out the parallel left shifter of one time. So, Steps (6a) through (6f) will be executed one time. After the first execution of Step (6a) is implemented, tube $T_0 = \varnothing$, tube $T_3 = \{A_{1,\,1}{}^0 A_{0,\,2}{}^1 A_{0,\,1}{}^1, A_{1,\,1}{}^0 A_{0,\,2}{}^0 A_{0,\,1}{}^1\}$, and tube $T_4 = \{A_{1,\,1}{}^0 A_{0,\,2}{}^1 A_{0,\,1}{}^0, A_{1,\,1}{}^0 A_{0,\,2}{}^0 A_{0,\,1}{}^0\}$. This is to say that the new execution environments for two bio-molecular deterministic one-tape Turing machines with the content of tape square, "$A_{0,\,1}{}^1$", and other two bio-molecular deterministic one-tape Turing machines with the content of tape square, "$A_{0,\,1}{}^0$" are, respectively, the fourth **BMPDTM** and the fifth **BMPDTM**. The position of the corresponding read-write head and the state of the corresponding finite state control are reserved. Figures 7.22 and 7.23 are employed to explain the result.

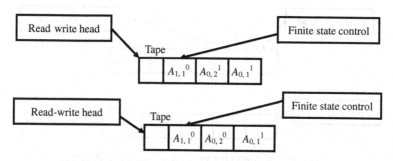

(a) The two **BMDTMs** in the fourth **BMPDTM**.

Fig. 7.22 Schematic representation of the current status of the execution environment to the fourth **BMPDTM**

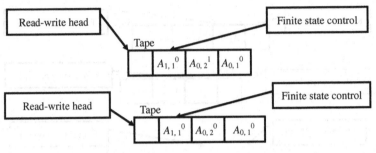

(a) The two **BMDTMs** in the fifth **BMPDTM**.

Fig. 7.23 Schematic representation of the current status of the execution environment to the fifth **BMPDTM**

Then, after the *first* execution of Step (6b) and the first execution of Step (6d) are finished, each operation returns a "yes" because the contents of those tubes are not empty. Therefore, from the first execution of Step (6c) and the first execution of Step (6e), $A_{1,2}^1$ and $A_{1,2}^0$ are, subsequently, appended into the head of each bit pattern in tube T_3 and the head of each bit pattern in tube T_4. This implies that the content of the fourth tape square for each tape of the two bio-molecular deterministic one-tape Turing machines in the fourth **BMPDTM** is written by the corresponding read-write head and is 1 ($A_{1,2} = 1$), and the content of the fourth tape square for each tape of the two bio-molecular deterministic one-tape Turing machines in the fifth **BMPDTM** is written by the corresponding read-write head and is 0 ($A_{1,2} = 0$). Simultaneously, the position of the corresponding read-write head is moved to the *left new* tape square, the state of the corresponding finite state control is changed as "$A_{1,2} = 1$" and "$A_{1,2} = 0$" and tube $T_3 = \{A_{1,2}^1 A_{1,1}^0 A_{0,2}^1 A_{0,1}^1, A_{1,2}^1 A_{1,1}^0 A_{0,2}^0 A_{0,1}^1\}$ and tube $T_4 = \{A_{1,2}^0 A_{1,1}^0 A_{0,2}^1 A_{0,1}^0, A_{1,2}^0 A_{1,1}^0 A_{0,2}^0 A_{0,1}^0\}$. Figures 7.24 and 7.25 are employed to illustrate the result.

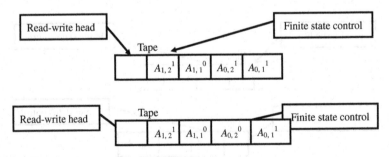

(a) The two **BMDTMs** in the fourth **BMPDTM**.

Fig. 7.24 Schematic representation of the current status of the execution environment to the fourth **BMPDTM**

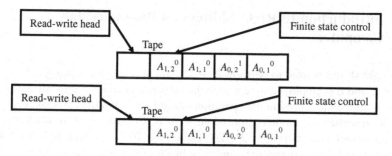

(a) The two **BMDTMs** in the fifth **BMPDTM**.

Fig. 7.25 Schematic representation of the current status of the execution environment to the fifth **BMPDTM**

Then, after the first execution of Step (6f) is finished, tube $T_0 = \{A_{1,2}{}^1 A_{1,1}{}^0 A_{0,}$ $_2{}^1 A_{0,1}{}^1, A_{1,2}{}^1 A_{1,1}{}^0 A_{0,2}{}^0 A_{0,1}{}^1, A_{1,2}{}^0 A_{1,1}{}^0 A_{0,2}{}^1 A_{0,1}{}^0, A_{1,2}{}^0 A_{1,1}{}^0 A_{0,2}{}^0 A_{0,1}{}^0\}$, tube $T_3 = \varnothing$ and tube $T_4 = \varnothing$. This is to say that the execution environment for the two bio-molecular deterministic one-tape Turing machines in the fourth **BMPDTM** and the two bio-molecular deterministic one-tape Turing machines in the fifth **BMPDTM** becomes the first **BMPDTM**. The position of the corresponding read-write head and the state of the corresponding finite state control are both reserved. Figure 7.26 is applied to explain the current status of the execution environment to the first **BMPDTM**. From Fig. 7.26, it is indicated that the contents to the four tapes in the execution environment of the first **BMPDTM** are not changed. Simultaneously, **Algorithm 7.3** is terminated.

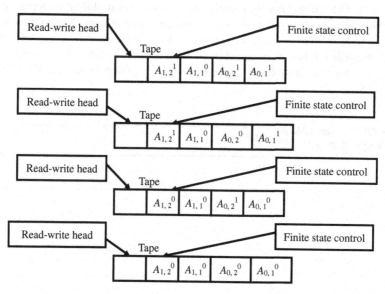

(a) The four **BMDTMs** in the first **BMPDTM**.

Fig.7.26 Schematic representation of the current status of the execution environment to the first **BMPDTM**

7.3 Introduction to Right Shifters on Bio-molecular Computer

The right shifter is used to perform $A \div 2^B$, where A and B are unsigned integers of n bits and B is employed to represent the number of bits for right shift. A symbol "$\gg$" is applied to represent the operation of right shift, and the expression $A \div 2^B$ can be rewritten as another expression: $A \gg B$. Consider how to carry out the computational task of $100 \div 2^2$. The expression $100 \div 2^2$ can be rewritten as $100 \gg 2$. This is to say that the number of bits for right shift is two. When the first execution of right shift for $100 \div 2^2$ is finished, the first bit, "0", is shifted out and is discarded, the second bit, "0" is shifted into the position of the first bit, the third bit, "1" is shifted into the position of the second bit, and a new bit, "0" is automatically put into the position of the third bit. Therefore, the intermediate result for the first right shift is 010. Next, when the second execution of left shift is implemented, the first bit, "0", is shifted out and is discarded, the second bit, "1" is shifted into the position of the first bit, the third bit, "0" is shifted into the position of the second bit, and a new bit, "0" is automatically put into the position of the third bit. Hence, the final result for the second right shift is 001. This is to say that the result for $100 \div 2^2$ is equal to 001.

In the processing of carrying out $A \div 2^B$, the *original* value for A is still represented as $A_{0, n}, A_{0, n-1}, \ldots, A_{0, 1}$ (denoted in Sect. 7.2), the intermediate value of the first right shift for A is still represented as $A_{1, n}, A_{1, n-1}, \ldots, A_{1, 1}$ (denoted in Sect. 7.2), the intermediate value of the dth right shift for A is still represented as $A_{d, n}, A_{d, n-1}, \ldots, A_{d, 1}$ (denoted in Sect. 7.2), and the final value of the *last* right shift for A is still represented as $A_{B, n}, A_{B, n-1}, \ldots, A_{B, 1}$ (denoted in Sect. 7.2). One truth table is usually used with a right shifter to represent all possible combinations of inputs and the corresponding outputs. The truth table for "$\gg$" (a right shifter) is shown in Table 7.8. Because the bit $A_{d-1, 1}$ is shifted out and is discarded, it is not contained in an input column in Table 7.8. Similarly, the bit $A_{d, n}$ is filled with a bit "0", so it is also not included in an output column in Table 7.8.

Table 7.8 The truth table for the dth right shift to $A \div 2^B$

Input $A_{d-1, k}$	Output $A_{d, k-1}$
0	0
1	1

7.3.1 The Construction for the Parallel Right Shifter on Bio-molecular Computer

For the sake of convenience, $A_{d,k}{}^1$ (denoted in Sect. 7.2.1) for $0 \le d \le B$ and $1 \le k \le n$ is that the value of $A_{d,k}$ is 1 and $A_{d,k}{}^0$ (denoted in Sect. 7.2.1) for $0 \le d \le B$ and $1 \le k \le n$ is that the value of $A_{d,k}$ is 0. The following algorithm is proposed to perform the parallel right shifter. Some notations in **Algorithm 7.4** are denoted in Sects. 7.2 and 7.2.1.

Algorithm 7.4: Parallel-Right-Shifter(T_0)

(1) Append-head($T_1, A_{0,n}{}^1$).

(2) Append-head($T_2, A_{0,n}{}^0$).

(3) $T_0 = \cup(T_1, T_2)$.

(4) **For** $k = n - 1$ **to** 1

    (4a) Amplify(T_0, T_1, T_2).

    (4b) Append($T_1, A_{0,k}{}^1$).

    (4c) Append($T_2, A_{0,k}{}^0$).

    (4d) $T_0 = \cup(T_1, T_2)$.

EndFor

(5) **For** $d = 1$ **to** B

    (5a) Append($T_0, A_{d,n}{}^0$).

  (6) **For** $k = n$ **downto** 2

    (6a) $T_3 = +(T_0, A_{d-1,k}{}^1)$ and $T_4 = -(T_0, A_{d-1,k}{}^1)$.

    (6b) **If** (Detect(T_3) == "yes") **Then**

        (6c) Append($T_3, A_{d,k-1}{}^1$).

    EndIf

    (6d) **If** (Detect(T_4) == "yes") **Then**

        (6e) Append($T_4, A_{d,k-1}{}^0$).

    EndIf

    (6f) $T_0 = \cup(T_3, T_4)$.

  EndFor

EndFor

End Algorithm

Lemma 7-4: *The algorithm, **Parallel-Right-Shifter**(T_0), can be employed to carry out the parallel right shifter.*

Proof The algorithm, **Parallel-Right-Shifter**(T_0), is implemented by means of the *extract, detect, append* and *merge* operations. On the execution for Steps (1) through (4d), they are mainly applied to generate solution space of 2^n unsigned integers of n bits for the only input (the range of values for them is from 0 to $2^n - 1$). After those operations are carried out, 2^n combinations of n bits are included in tube T_0.

Step (5) is a nested loop and is used to finish the parallel right shift of B times for 2^n combinations of n bits in tube T_0. Each execution of Step (5a) applies the *append* operation to put the value "0" into the position of the nth bit to each right shift. Step (6) is the inner loop and is mainly used to perform the right shift of one time. From each execution of Step (6a), it uses the *extract* operation to generate two different tubes containing different inputs. This implies T_3 includes all of the inputs that have $A_{d-1, k} = 1$ and T_4 contains all of the inputs that have $A_{d-1, k} = 0$. Next, each execution of Step (6b) and each execution of Step (6d) applies the *detect* operations to check whether consists of any input for tubes T_3 and T_4 or not. If a "yes" is returned from Step (6b), then each execution of Step (6c) uses the *append* operation to put the value "1" into the position of the $(k-1)$th bit $(A_{d, k-1} = 1)$ in tube T_3 for the right shift of the dth time. Similarly, if a "yes" is returned from Step (6d), then each execution of Step (6e) applies the *append* operation to put the value "0" into the position of the $(k-1)$th bit $(A_{d, k-1} = 0)$ in tube T_4 for the right shift of the dth time. Then, each execution of Step (6f) uses the *merge* operation to pour tubes T_3 and T_4 into tube T_0. Repeat execution of each operation is the nested loop until right shift the last time is performed. Each input in tube T_0 performs its value $\div 2^B$. ∎

7.3.2 The Power for the Parallel Right Shifters of N Bits on Bio-molecular Computer

Consider that four values for an unsigned integer of two bits are, subsequently, $00(0_{10}) (A_{0, 2}{}^0 A_{0, 1}{}^0)$, $01(1_{10}) (A_{0, 2}{}^0 A_{0, 1}{}^1)$, $10(2_{10}) (A_{0, 2}{}^1 A_{0, 1}{}^0)$ and $11(3_{10}) (A_{0, 2}{}^1 A_{0, 1}{}^1)$. For any given value 2^1, we want to simultaneously perform $(A_{0, 2}{}^0 A_{0, 1}{}^0) \div 2^1$, $(A_{0, 2}{}^0 A_{0, 1}{}^1) \div 2^1$, $(A_{0, 2}{}^1 A_{0, 1}{}^0) \div 2^1$, and $(A_{0, 2}{}^1 A_{0, 1}{}^1) \div 2^1$. **Algorithm 7.4**, **Parallel-Right-Shifter**(T_0), can be used to perform the computational task. Tube T_0 is an empty tube and is regarded as an input tube of **Algorithm 7.4**. In light of Definition 5−2, the input tube T_0 is regarded as the execution environment of the first **BMPDTM**. Similarly, tubes T_1, T_2, T_3 and T_4 used in **Algorithm 7.4** also are regarded, subsequently, as the execution environment of the second **BMPDTM**, the execution environment of the third **BMPDTM**, the execution environment of the fourth **BMPDTM**, and the execution environment of the fifth **BMPDTM**.

Steps (1) through (4d) in **Algorithm 7.4** are used to generate a **BMPDTM** with four bio-molecular deterministic one-tape Turing machines. After the execution for Step (1) and Step (2) of **Algorithm 7.4** is implemented, since the number of bits for representing those four values is two, tube $T_1 = \{A_{0, 2}{}^1\}$ and tube $T_2 = \{A_{0, 2}{}^0\}$. This indicates that a **BMDTM** in the second **BMPDTM** and in the third **BMPDTM** is generated. Figure 7.27 is applied to show the current status of the execution environment to the second **BMPDTM** and the third **BMPDTM**. From Fig. 7.27, the content of the first tape square for the tape in the first **BMDTM** in the second **BMPDTM** is written by its corresponding read-write head and is 1 $(A_{0, 2} = 1)$, and the content of the first tape square for the tape in the first **BMDTM** in the third **BMPDTM** is written by its corresponding read-write head and is 0 ($A_{0,\,2} = 0$). For the first **BMDTM** in the second **BMPDTM**, the position of the read-write head is moved to the *right new* tape square, and the state of the finite state control is changed as "$A_{0,\,2} = 1$". Similarly, for the first **BMDTM** in the third **BMPDTM**, the position of the read-write head is moved to the *right new* tape square, and the state of the finite state control is changed as "$A_{0,\,2} = 0$".

Next, after the execution for Step (3) of **Algorithm 7.4** is performed, tube $T_0 = \{A_{0,\,2}{}^1,\ A_{0,\,2}{}^0\}$, tube $T_1 = \varnothing$ and tube $T_2 = \varnothing$. This is to say that the execution environment for the first bio-molecular deterministic one-tape Turing machine in the second **BMPDTM** and the first bio-molecular deterministic one-tape Turing machine in the third **BMPDTM** becomes the first **BMPDTM**. The position of the corresponding read-write head and the state of the corresponding finite state control are both reserved. Figure 7.28 is employed to explain the current status of the execution environment to the first **BMPDTM**. From Fig. 7.28, it is pointed out that the contents to the two tapes in the execution environment of the first **BMPDTM** are not changed.

Step (4) is the first loop in **Algorithm 7.4**, since the number of bits for representing those four values is two, the lower bound ($n - 1$) is one. Therefore, after

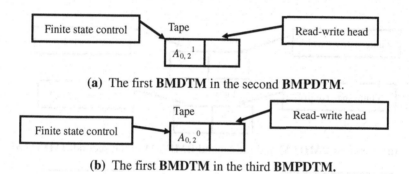

(a) The first **BMDTM** in the second **BMPDTM**.

(b) The first **BMDTM** in the third **BMPDTM**.

Fig. 7.27 Schematic representation of the current status of the execution environment to the second **BMPDTM** and the third **BMPDTM**

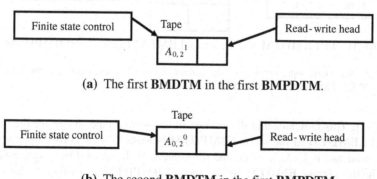

(a) The first **BMDTM** in the first **BMPDTM**.

(b) The second **BMDTM** in the first **BMPDTM**.

Fig. 7.28 Schematic representation of the current status of the execution environment to the first **BMPDTM**

the first execution of Step (4a) is performed, tube $T_0 = \varnothing$, tube $T_1 = \{A_{0,\,2}{}^1, A_{0,\,2}{}^0\}$ and tube $T_2 = \{A_{0,\,2}{}^1, A_{0,\,2}{}^0\}$. This implies that the first **BMDTM** and the second **BMDTM** in the execution environment of the first **BMPDTM** are both copied into the second **BMPDTM** and the third **BMPDTM**. Figure 7.29 is applied to reveal the current status of the execution environment to the second **BMPDTM** and the third **BMPDTM**. From Fig. 7.29, the contents of the first tape square for the corresponding tape of the first **BMDTM** and the corresponding tape of the second **BMDTM** in the execution environment of the second **BMPDTM** are, subsequently, 1 ($A_{0,\,2} = 1$) and 0 ($A_{0,\,2} = 0$). The contents of the first tape square for the corresponding tape of the first **BMDTM** and the corresponding tape of the second **BMDTM** in the execution environment of the third **BMPDTM** are also, respectively, 0 ($A_{0,\,2} = 0$) and 1 ($A_{0,\,2} = 1$). From Fig. 7.29, four bio-molecular deterministic one-tape Turing machines are produced. For the four bio-molecular deterministic one-tape Turing machines, the position of the corresponding read-write head and the state of the corresponding finite state control are reserved.

Next, after the first execution for Step (4b) and Step (4c) of **Algorithm 7.4** is performed, tube $T_1 = \{A_{0,\,2}{}^1 A_{0,\,1}{}^1, A_{0,\,2}{}^0 A_{0,\,1}{}^1\}$ and tube $T_2 = \{A_{0,\,2}{}^0 A_{0,\,1}{}^0, A_{0,\,}$

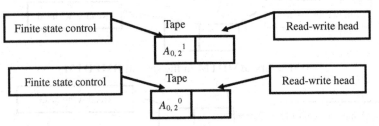

(a) The first **BMDTM** and the second **BMDTM** in the second **BMPDTM**.

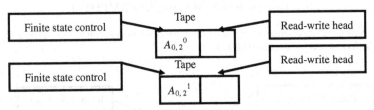

(b) The first **BMDTM** and the second **BMDTM** in the third **BMPDTM**.

Fig. 7.29 Schematic representation of the current status of the execution environment to the second **BMPDTM** and the third **BMPDTM**

$_2{}^1 A_{0,\,1}{}^0\}$. This indicates that the content of the *second* tape square for the tape in the first **BMDTM** in the second **BMPDTM** is written by its corresponding read-write head and is 1 ($A_{0,\,1} = 1$), and the content of the *second* tape square for the tape in the second **BMDTM** in the second **BMPDTM** is written by its corresponding read-write head and is also 1 ($A_{0,\,1} = 1$). Simultaneously, the position of the corresponding read-write head is both moved to the *right new* tape square and the state of the corresponding finite state control is both changed as "$A_{0,1} = 1$". Similarly, the content of the *second* tape square for the tape in the first **BMDTM** in the third **BMPDTM** is written by its corresponding read-write head and is 0 ($A_{0,1} = 0$), and the content of the *second* tape square for the tape in the second **BMDTM** in the third **BMPDTM** is written by its corresponding read-write head and is also 0 ($A_{0,1} = 0$). Simultaneously, the position of the corresponding read-write head is both moved to the *right new* tape square and the state of the corresponding finite state control is both changed as "$A_{0,1} = 0$". Figure 7.30 is used to illustrate the current status of the execution environment to the second **BMPDTM** and the third **BMPDTM**.

Next, after the first execution for Step (4d) of **Algorithm 7.4** is implemented, tube $T_0 = \{A_{0,2}{}^1 A_{0,1}{}^1, A_{0,2}{}^1 A_{0,1}{}^0, A_{0,2}{}^0 A_{0,1}{}^1, A_{0,2}{}^0 A_{0,1}{}^0\}$, tube $T_1 = \varnothing$ and

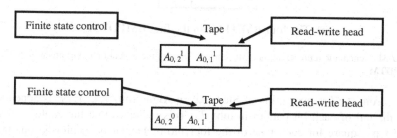

(a) The first **BMDTM** and the second **BMDTM** in the second **BMPDTM**.

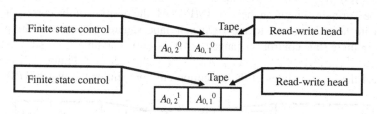

(b) The first **BMDTM** and the second **BMDTM** in the third **BMPDTM**.

Fig. 7.30 Schematic representation of the current status of the execution environment to the second **BMPDTM** and the third **BMPDTM**

tube $T_2 = \varnothing$. This implies that the execution environment for those bio-molecular deterministic one-tape Turing machines in the second **BMPDTM** and in the third **BMPDTM** becomes the first **BMPDTM**. Figure 7.31 is employed to reveal the current status of the execution environment to the first **BMPDTM**. From Fig. 7.31, the contents to the four tapes in the execution environment of the first **BMPDTM** are not changed, and the position of the corresponding read-write head and the state of the corresponding finite state control are reserved.

Next, because Step (5) of **Algorithm 7.4** is the nested loop and the upper bound for the outer loop in Step (5) of **Algorithm 7.4** is 1, Step (5a) will be executed one

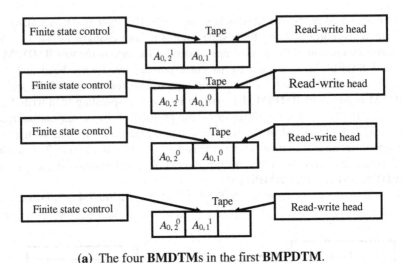

(a) The four **BMDTMs** in the first **BMPDTM**.

Fig. 7.31 Schematic representation of the current status of the execution environment to the first **BMPDTM**

time. From the first execution of Step (5a) in **Algorithm 7.4**, $A_{1, 2}^{0}$ is appended into the tail of each bit pattern in tube T_0. This indicates that the content of the third tape square for each tape of the four bio-molecular deterministic one-tape Turing machines in the first **BMPDTM** is written by the corresponding read-write head and is 0 ($A_{1, 2} = 0$). Simultaneously, for the four bio-molecular deterministic one-tape Turing machines in the first **BMPDTM**, the position of the corresponding read-write head is moved to the *right new* tape square, the state of the corresponding finite state control is changed as "$A_{1, 2} = 0$" and tube $T_0 = \{A_{0, 2}^{1} A_{0, 1}^{1} A_{1, 2}^{0}, A_{0, 2}^{1} A_{0, 1}^{0} A_{1, 2}^{0}, A_{0, 2}^{0} A_{0, 1}^{1} A_{1, 2}^{0}, A_{0, 2}^{0} A_{0, 1}^{0} A_{1, 2}^{0}\}$. Figure 7.32 is used to reveal the current status of the execution environment to the first **BMPDTM**.

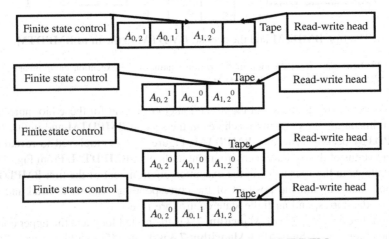

(a) The four **BMDTMs** in the first **BMPDTM**.

Fig. 7.32 Schematic representation of the current status of the execution environment to the first **BMPDTM**

Step (6) in **Algorithm 7.3** is the inner loop, its lower bound and upper bound are two, and is employed to perform the parallel right shifter of one time. So, Steps (6a) through (6f) will be executed one time. After the first execution of Step (6a) is implemented, tube $T_0 = \varnothing$, tube $T_3 = \{A_{0,\,2}{}^1 A_{0,\,1}{}^1 A_{1,\,2}{}^0, A_{0,\,2}{}^1 A_{0,\,1}{}^0 A_{1,\,2}{}^0\}$, and tube $T_4 = \{A_{0,\,2}{}^0 A_{0,\,1}{}^1 A_{1,\,2}{}^0, A_{0,\,2}{}^0 A_{0,\,1}{}^0 A_{1,\,2}{}^0\}$. This indicates that the new execution environments for two bio-molecular deterministic one-tape Turing machines with the content of tape square, "$A_{0,\,2}{}^1$", and other two bio-molecular deterministic one-tape Turing machines with the content of tape square, "$A_{0,\,2}{}^0$" are, respectively, the fourth **BMPDTM** and the fifth **BMPDTM**. The position of the corresponding read-write head and the state of the corresponding finite state control are reserved. Figures 7.33 and 7.34 are used to illustrate the result.

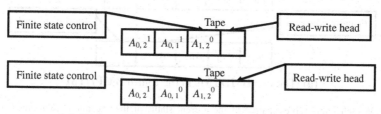

(a) The two **BMDTMs** in the fourth **BMPDTM**.

Fig. 7.33 Schematic representation of the current status of the execution environment to the fourth **BMPDTM**

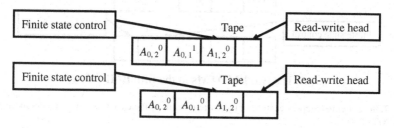

(a) The two **BMDTMs** in the fifth **BMPDTM**.

Fig. 7.34 Schematic representation of the current status of the execution environment to the fifth **BMPDTM**

Then, after the *first* execution of Step (6b) and the first execution Step (6d) are carried out, each operation returns a "yes" because the contents of those tubes are not empty. Thus, from the first execution of Step (6c) and the first execution of Step (6e), $A_{1,\,1}{}^1$ and $A_{1,\,1}{}^0$ are, subsequently, appended into the tail of each bit pattern in tube T_3 and the tail of each bit pattern in tube T_4. This is to say that the content of the fourth tape square for each tape of the two bio-molecular deterministic one-tape Turing machines in the fourth **BMPDTM** is written by the corresponding read-write head and is 1 ($A_{1,\,1} = 1$), and the content of the fourth tape square for each tape of the two bio-molecular deterministic one-tape Turing machines in the fifth **BMPDTM** is written by the corresponding read-write head and is 0 ($A_{1,\,1} = 0$). Simultaneously, the position of the corresponding read-write head is moved to the *right new* tape square, the state of the corresponding finite state control is changed as "$A_{1,\,1} = 1$" and "$A_{1,\,1} = 0$" and tube $T_3 = \{A_{0,\,2}{}^1 A_{0,\,1}{}^1 A_{1,\,2}{}^0 A_{1,\,1}{}^1, A_{0,\,2}{}^1 A_{0,\,1}{}^0 A_{1,\,2}{}^0 A_{1,\,1}{}^1\}$ and tube $T_4 = \{A_{0,\,2}{}^0 A_{0,\,1}{}^1 A_{1,\,2}{}^0 A_{1,\,1}{}^0, A_{0,\,2}{}^0 A_{0,\,1}{}^0 A_{1,\,2}{}^0 A_{1,\,1}{}^0\}$. Figures 7.35 and 7.36 are used to reveal the result.

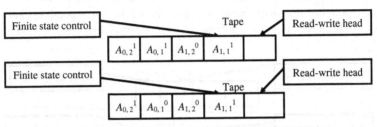

(a) The two **BMDTMs** in the fourth **BMPDTM**.

Fig. 7.35 Schematic representation of the current status of the execution environment to the fourth **BMPDTM**

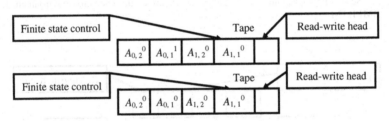

(a) The two **BMDTMs** in the fifth **BMPDTM**.

Fig. 7.36 Schematic representation of the current status of the execution environment to the fifth **BMPDTM**

Then, after the first execution of Step (6f) is finished, tube $T_0 = \{A_{0,\,2}{}^1 A_{0,\,1}{}^1 A_{1,\,2}{}^0 A_{1,\,1}{}^1, A_{0,\,2}{}^1 A_{0,\,1}{}^0 A_{1,\,2}{}^0 A_{1,\,1}{}^1, A_{0,\,2}{}^0 A_{0,\,1}{}^1 A_{1,\,2}{}^0 A_{1,\,1}{}^0, A_{0,\,2}{}^0 A_{0,\,1}{}^0 A_{1,\,2}{}^0 A_{1,\,1}{}^0\}$, tube $T_3 = \varnothing$ and tube $T_4 = \varnothing$. This implies that the execution environment for the two bio-molecular deterministic one-tape Turing machines in the fourth **BMPDTM** and the two bio-molecular deterministic one-tape Turing machines in the fifth **BMPDTM** becomes the first **BMPDTM**. The position of the corresponding read-write head and the state of the corresponding finite state control are both reserved. Figure 7.37 is applied to explain the current status of the execution environment to the first **BMPDTM**. From Fig. 7.37, it is indicated that the contents to the four tapes in the execution environment of the first **BMPDTM** are not changed. Simultaneously, **Algorithm 7.4** is terminated.

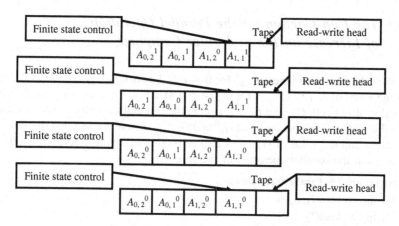

(a) The four **BMDTM**s in the first **BMPDTM**.

Fig. 7.37 Schematic representation of the current status of the execution environment to the first **BMPDTM**

7.4 Introduction to the Increase Operation on Bio-molecular Computer

The increase operation is applied to carry out "$I = I + 1$", where I is unsigned integers of n bits. A symbol "++" is used to represent the operation of increase, and the expression, $I = I + 1$, can be rewritten as another expression: I ++. An unsigned integer of n bits, I, is regarded as the augend and the sum in the expression I ++. The value "1" is also regarded as the addend in the expression I ++. Suppose that I can be represented as $I_{e,\, n}, I_{e,\, n-1}, \ldots, I_{e\, 1}$ for $0 \le e \le 1$, where the value for each $I_{e,\, k}$ for $1 \le k \le n$ is 1 or 0. Also assume that two binary numbers, G_k and G_{k-1} for $1 \le k \le n$, are applied to respectively represent the carry of the augend bit and the addend bit and the previous carry.

In the processing of performing I ++, suppose that the *original* value for I is represented as $I_{0,\, n}, I_{0,\, n-1}, \ldots, I_{0,\, 1}$, and the sum to I ++ is represented as $I_{1,\, n}, I_{1,\, n-1}, \ldots, I_{1,\, 1}$. Table 7.9 is regarded as the truth table of a one-bit adder for $I_{0,\, 1}$, G_0, $I_{1,\, 1}$ and G_1, and Table 7.10 is also regarded as the truth table of a one-bit adder for $I_{0,\, k}$, G_{k-1}, $I_{1,\, k}$, and G_k for $2 \le k \le n$.

Table 7.9 The truth table of a one-bit adder for $I_{0,\, 1}, G_0,$ $I_{1,\, 1}$ and G_1

$I_{0,\, 1}$	The first bit of the addend	G_0	$I_{1,\, 1}$	G_1
0	1	0	1	0
1	1	0	0	1

Table 7.10 The truth table of a one-bit adder for $I_{0,\, k}$, $G_{k-1}, I_{1,\, k},$ and G_k for $2 \le k \le n$

$I_{0,\, k}$	The kth bit of the addend	G_{k-1}	$I_{1,\, k}$	G_k
0	0	0	0	0
0	0	1	1	0
1	0	0	1	0
1	0	1	0	1

7.4.1 The Construction for the Parallel Operation of Increase on Bio-molecular Computer

For the sake of convenience, $I_{e,k}{}^1$ for $0 \le e \le 1$ and $1 \le k \le n$ denotes the fact that the value of $I_{e,k}$ is 1, $I_{e,k}{}^0$ for $0 \le e \le 1$ and $1 \le k \le n$ denotes the fact that the value of $I_{e,k}$ is 0, $G_k{}^1$ and $G_{k-1}{}^1$ denote the fact that the value of G_k is one and the value of G_{k-1} is also one, and $G_k{}^0$ and $G_{k-1}{}^0$ denote the fact that the value of G_k is zero and the value of G_{k-1} is also zero. The following algorithm is offered to carry out the parallel operation of increase.

Algorithm 7.5: ParallelIncrease(T_0)

(1) Append-head($T_1, I_{0,1}{}^1$).

(2) Append-head($T_2, I_{0,1}{}^0$).

(3) $T_0 = (T_1, T_2)$.

(4) **For** $k = 2$ **to** n

      (4a) Amplify(T_0, T_1, T_2).

      (4b) Append-head($T_1, I_{0,k}{}^1$).

      (4c) Append-head($T_2, I_{0,k}{}^0$).

      (4d) $T_0 = (T_1, T_2)$.

EndFor

(5) Append-head($T_0, G_0{}^0$).

(6) $T_3 = +(T_0, I_{0,1}{}^1)$ and $T_4 = (T_0, I_{0,1}{}^1)$.

(7) **If** (Detect(T_3) $==$ "yes") **Then**

      (7a) Append-head($T_3, I_{1,1}{}^0$) and Append-head($T_3, G_1{}^1$).

  EndIf

(8) **If** (Detect(T_4) $==$ "yes") **Then**

      (8a) Append-head($T_4, I_{1,1}{}^1$) and Append-head($T_4, G_1{}^0$).

  EndIf

(9) $T_0 = (T_3, T_4)$.

(10) **For** $k = 2$ **to** n

      (10a) $T_5 = +(T_0, I_{0,k}{}^1)$ and $T_6 = (T_0, I_{0,k}{}^1)$.

      (10b) $T_7 = +(T_5, G_{k-1}{}^1)$ and $T_8 = (T_5, G_{k-1}{}^1)$.

      (10c) $T_9 = +(T_6, G_{k-1}{}^1)$ and $T_{10} = (T_6, G_{k-1}{}^1)$.

      (10d) **If** (Detect(T_7) $==$ "yes") **Then**

          (10e) Append-head($T_7, I_{1,k}{}^0$) and Append-head($T_7, G_k{}^1$).

      EndIf

      (10f) **If** (Detect(T_8) $==$ "yes") **Then**

          (10g) Append-head($T_8, I_{1,k}{}^1$) and Append-head($T_8, G_k{}^0$).

      EndIf

      (10h) **If** (Detect(T_9) $==$ "yes") **Then**

          (10i) Append-head($T_9, I_{1,k}{}^1$) and Append-head($T_9, G_k{}^0$).

      EndIf

      (10j) **If** (Detect(T_{10}) $==$ "yes") **Then**

          (10k) Append-head($T_{10}, I_{1,k}{}^0$) and Append-head($T_{10}, G_k{}^0$).

      EndIf

(10l) $T_0 =$ (T_7, T_8, T_9, T_{10}).

EndFor

EndAlgorithm

Lemma 7-5: *The algorithm, **ParallelIncrease**(T_0), can be used to perform the parallel operation of increase.*

Proof The algorithm, **ParallelIncrease**(T_0), is implemented by means of the *extract*, *amplify*, *detect*, *append-head* and *merge* operations. From each execution for Steps (1) through (4d), solution space of 2^n unsigned integers of n bits (the range of values for them is from 0 to $2^n - 1$) is generated. After those operations are performed, 2^n combinations of n bits are contained in tube T_0. Because the previous carry for addition of the first bit to each augend and addend is zero, from the execution of Step (5), the value "0" of G_0 is appended into the head of each bit pattern in tube T_0.

From each execution of Step (6), tube T_3 contains all of the inputs that have $I_{0, 1} = 1$, representing the first column of the second row in Table 7.9, and tube T_4 consists of all of the inputs that have $I_{0, 1} = 0$, representing the first column of the first row in Table 7.9. Since the contents for tubes T_3 and T_4 are not empty, a "yes" is returned from Step (7) and Step (8). Therefore, from each execution of Step (7a), the value "0" of $I_{1, 1}$ and the value "1" of G_1 are appended into the head of each bit pattern in tube T_3, and from each execution of Step (8a), the value "1" of $I_{1, 1}$ and the value "0" of G_1 are appended into the head of each bit pattern in tube T_4. This implies that Table 7.9 is performed. Next, each execution of Step (9) applies the *merge* operation to pour tubes T_3 through T_4 into tube T_0. Tube T_0 contains the result performing Table 7.9.

Step (10) is the second loop and is used to finish the parallel one-bit adder of $(n - 1)$ times. From each execution for Steps (10a) through (10c), T_5 includes all of the inputs that have $I_{0, k} = 1$, T_6 contains all of the inputs that have $I_{0, k} = 0$, T_7 consists of all of the inputs that have $I_{0, k} = 1$ and $G_k = 1$, T_8 includes all of the inputs that have $I_{0, k} = 1$ and $G_k = 0$, T_9 consists of all of the inputs that have $I_{0, k} = 0$ and $G_k = 1$, and T_{10} includes all of the inputs that have $I_{0, k} = 0$ and $G_k = 0$. Having performed Steps (10a) through (10c), this is to say that four different inputs of a one-bit adder as shown in Table 7.10 were poured into tubes T_7 through T_{10}, respectively.

From each execution for Steps (10d), (10f), (10h) and (10j), because the contents for tubes T_7, T_8, T_9 and T_{10} are all not empty, therefore, a "yes" is returned from each step. Next, on each execution for Steps (10e), (10g), (10i) and (10k), the *append-head* operations are applied to append $I_{1, k}{}^1$ or $I_{1, k}{}^0$, and $G_k{}^1$ or $G_k{}^0$ onto the head of every bit pattern in the corresponding tubes. After performing Steps (10a) through (10k), we can say that four different outputs of a one-bit adder in Table 7.10 are appended into tubes T_7 through T_{10}. Next, each execution of Step (10l) applies the *merge* operation to pour tubes T_7 through T_{10} into tube T_0. Tube T_0 contains the result performing Table 7.10. Repeat execution of Steps (10a) through (10l) until the most significant bit for the augend and the addend is processed. Tube T_0 obtains the result performing the parallel operation of increase for 2^n unsigned integers of n bits. ∎

7.4.2 The Power for the Parallel Operation of Increase on Bio-molecular Computer

Consider that four values for an unsigned integer of two bits are, respectively, $00(0_{10})$ $(I_{0, 2}{}^0 I_{0, 1}{}^0)$, $01(1_{10})$ $(I_{0, 2}{}^0 I_{0, 1}{}^1)$, $10(2_{10})$ $(I_{0, 2}{}^1 I_{0, 1}{}^0)$, and $11(3_{10})$ $(I_{0, 2}{}^1 I_{0, 1}{}^1)$. We want to simultaneously increase each value of the four values. **Algorithm 7.5**, **ParallelIncrease**(T_0), can be used to perform the computational task. Tube T_0 is an empty tube and is regarded as an input tube of **Algorithm 7.5**. Due to Definition 5−2, the input tube T_0 can be regarded as the execution environment of the first **BMPDTM**. Similarly, each tube T_k in **Algorithm 7.5** for $1 \le k \le 10$ can be also regarded as the execution environment of the $(k + 1)$th **BMPDTM**.

Steps (1) through (4d) in **Algorithm 7.5** are applied to construct a **BMPDTM** with four bio-molecular deterministic one-tape Turing machines. After the first execution for Step (1) and Step (2) is performed, tube $T_1 = \{I_{0, 1}{}^1\}$ and tube $T_2 = \{I_{0, 1}{}^0\}$. This implies that a **BMDTM** in the second **BMPDTM** and in the third **BMPDTM** is generated. Figure 7.38 is employed to illustrate the current status of the execution environment to the second **BMPDTM** and the third **BMPDTM**. From Fig. 7.38, the content of the first tape square for the tape in the first **BMDTM** in the second **BMPDTM** is written by its corresponding read-write head and is 1 $(I_{0, 1} = 1)$, and the content of the first tape square for the tape in the first **BMDTM** in the third **BMPDTM** is written by its corresponding read-write head and is 0 $(I_{0, 1} = 0)$. Simultaneously, for the two **BMDTMs**, the position of the corresponding read-write head is moved to the *left new* tape square, and the status of the corresponding finite state control is, respectively, "$I_{0, 1} = 1$" and "$I_{0, 1} = 0$".

Next, after the execution for Step (3) is finished, tube $T_0 = \{I_{0, 1}{}^1, I_{0, 1}{}^0\}$, tube $T_1 = \varnothing$ and tube $T_2 = \varnothing$. This indicates that the execution environment for the

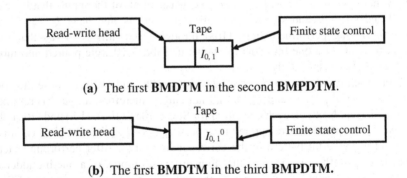

(a) The first **BMDTM** in the second **BMPDTM**.

(b) The first **BMDTM** in the third **BMPDTM**.

Fig. 7.38 Schematic representation of the current status of the execution environment to the second **BMPDTM** and the third **BMPDTM**

first bio-molecular deterministic one-tape Turing machine in the second **BMPDTM** and the first **BMDTM** in the third **BMPDTM** becomes the first **BMPDTM**. From Fig. 7.39, the contents to the two tapes in the execution environment of the first **BMPDTM** are not changed, and the position of each read-write head and the status of each finite state control are reserved.

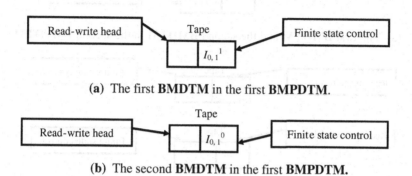

(a) The first **BMDTM** in the first **BMPDTM**.

(b) The second **BMDTM** in the first **BMPDTM**.

Fig. 7.39 Schematic representation of the current status of the execution environment to the first **BMPDTM**

Step (4) is the first loop and the upper bound (n) is two because the number of bits for representing those four values is two. Therefore, after the first execution of Step (4a) is implemented, tube $T_0 = \varnothing$, tube $T_1 = \{I_{0,\,1}{}^1, I_{0,\,1}{}^0\}$ and tube $T_2 = \{I_{0,\,1}{}^1, I_{0,\,1}{}^0\}$. This implies that the first **BMDTM** and the second **BMDTM** in the execution environment of the first **BMPDTM** are both copied into the second **BMPDTM** and the third **BMPDTM**. Figure 7.40 is used to reveal the current status of the execution environment to the second **BMPDTM** and the third **BMPDTM**. From Fig. 7.40, the contents of the first tape square for the corresponding tape of the first **BMDTM** and the corresponding tape of the second **BMDTM** in the execution environment of the second **BMPDTM** are, respectively, 1 ($I_{0,\,1} = 1$) and 0 ($I_{0,\,1} = 0$). The contents of the first tape square for the corresponding tape of the first **BMDTM** and the corresponding tape of the second **BMDTM** in the execution environment of the third **BMPDTM** are also, respectively, 0 ($I_{0,\,1} = 0$) and 1 ($I_{0,\,1} = 1$). From Fig. 7.40, it is indicated that four bio-molecular deterministic one-tape Turing machines are generated.

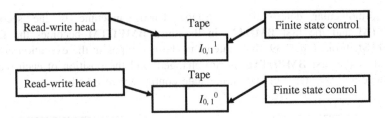

(a) The first **BMDTM** and the second **BMDTM** in the second **BMPDTM**.

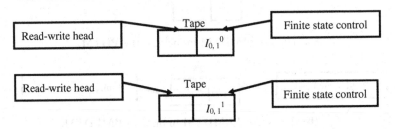

(b) The first **BMDTM** and the second **BMDTM** in the third **BMPDTM**.

Fig. 7.40 Schematic representation of the current status of the execution environment to the second **BMPDTM** and the third **BMPDTM**

Next, after the first execution for Step (4b) and Step (4c) is performed, tube $T_1 = \{I_{0, 2}{}^1 I_{0, 1}{}^1, I_{0, 2}{}^1 I_{0, 1}{}^0\}$ and tube $T_2 = \{I_{0, 2}{}^0 I_{0, 1}{}^1, I_{0, 2}{}^0 I_{0, 1}{}^0\}$. This is to say that the content of the *second* tape square for the tape in the first **BMDTM** in the second **BMPDTM** is written by its corresponding read-write head and is 1 ($I_{0, 2} = 1$), and the content of the *second* tape square for the tape in the second **BMDTM** in the second **BMPDTM** is written by its corresponding read-write head and is also 1 ($I_{0, 2} = 1$). Similarly, the content of the *second* tape square for the tape in the first **BMDTM** in the third **BMPDTM** is written by its corresponding read-write head and is 0 ($I_{0, 2} = 0$), and the content of the *second* tape square for the tape in the second **BMDTM** in the third **BMPDTM** is written by its corresponding read-write head and is also 0 ($I_{0, 2} = 0$). Figure 7.41 is used to illustrate the current status of the execution environment to the second **BMPDTM** and the third **BMPDTM**. From Fig. 7.41, the position of each read-write head is moved to the left new tape square, and the status of each finite state control is changed as "$I_{0, 2}{}^1$" or "$I_{0, 2}{}^0$".

Next, after the first execution for Step (4d) is implemented, tube $T_0 = \{I_{0, 2}{}^1 I_{0, 1}{}^1, I_{0, 2}{}^1 I_{0, 1}{}^0, I_{0, 2}{}^0 I_{0, 1}{}^1, I_{0, 2}{}^0 I_{0, 1}{}^0\}$, tube $T_1 = \varnothing$ and tube $T_2 = \varnothing$. This implies that the execution environment for those bio-molecular deterministic one-tape

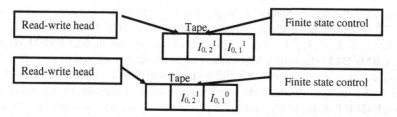

(a) The first **BMDTM** and the second **BMDTM** in the second **BMPDTM**.

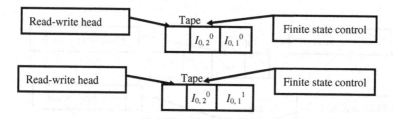

(b) The first **BMDTM** and the second **BMDTM** in the third **BMPDTM**.

Fig. 7.41 Schematic representation of the current status of the execution environment to the second **BMPDTM** and the third **BMPDTM**

Turing machines in the second **BMPDTM** and in the third **BMPDTM** becomes the first **BMPDTM**. Figure 7.42 is employed to reveal the current status of the execution environment to the first **BMPDTM**. From Fig. 7.42, the contents to the four tapes in the execution environment of the first **BMPDTM** are not changed, and the position of the corresponding read-write head and the state of the corresponding finite state control are reserved.

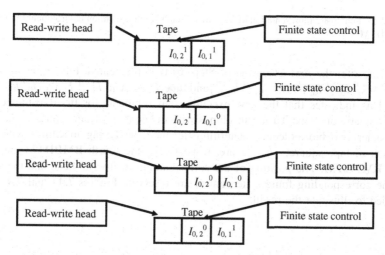

(a) The four **BMDTMs** in the first **BMPDTM**.

Fig. 7.42 Schematic representation of the current status of the execution environment to the first **BMPDTM**

Next, after the first execution of Step (5) is implemented, from each read-write head, $G_0{}^0$ is written into each tape. Therefore, tube $T_0 = \{G_0{}^0 I_{0,\,2}{}^1 I_{0,\,1}{}^1, G_0{}^0 I_{0,\,2}{}^1 I_{0,\,1}{}^0, G_0{}^0 I_{0,\,2}{}^0 I_{0,\,1}{}^1, G_0{}^0 I_{0,\,2}{}^0 I_{0,\,1}{}^0\}$. This is to say that for the four **BMDTMs** in the first **BMPDTM** the position of each read-write head is moved to the *left new* tape square and the status of each finite state control is changed as "$G_0 = 0$". Figure 7.43 is used to illustrate the current status of the execution environment to the first **BMPDTM**. From Fig. 7.43, the position of each read-write head is moved to the left new tape square, and the position of each finite state control is changed as "$G_0 = 0$".

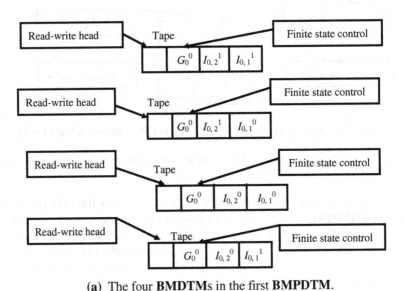

(a) The four **BMDTMs** in the first **BMPDTM**.

Fig. 7.43 Schematic representation of the current status of the execution environment to the first **BMPDTM**

Next, after the *first* execution of Step (6) is implemented, tube $T_0 = \varnothing$, tube $T_3 = \{G_0{}^0 I_{0,\,2}{}^1 I_{0,\,1}{}^1, G_0{}^0 I_{0,\,2}{}^0 I_{0,\,1}{}^1\}$ and tube $T_4 = \{G_0{}^0 I_{0,\,2}{}^1 I_{0,\,1}{}^0, G_0{}^0 I_{0,\,2}{}^0 I_{0,\,1}{}^0\}$. This indicates that the new execution environments for two bio-molecular deterministic one-tape Turing machines with the content of tape square, "$I_{0,\,1}{}^1$", and other two bio-molecular deterministic one-tape Turing machines with the content of tape square, "$I_{0,\,1}{}^0$" are, respectively, the fourth **BMPDTM** and the fifth **BMPDTM**. The position of the corresponding read-write head and the state of the corresponding finite state control are reserved. Figures 7.44 and 7.45 are applied to illustrate the result.

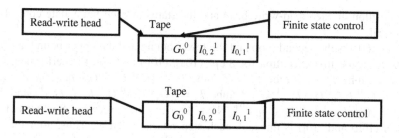

(a) The two **BMDTMs** in the fourth **BMPDTM**.

Fig. 7.44 Schematic representation of the current status of the execution environment to the fourth **BMPDTM**

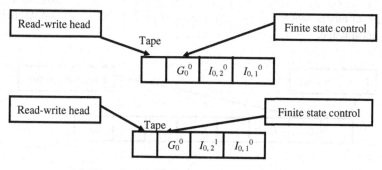

(a) The two **BMDTMs** in the fifth **BMPDTM**.

Fig. 7.45 Schematic representation of the current status of the execution environment to the fifth **BMPDTM**

Because tube $T_3 \neq \varnothing$ and tube $T_4 \neq \varnothing$, from the first execution of Step (7) and the first execution of Step (8), the conditions are true. Therefore, after the first execution for Steps (7a) and (8a) is performed, tube $T_3 = \{G_1{}^1 I_{1,1}{}^0 G_0{}^0 I_{0,2}{}^1 I_{0,1}{}^1, G_1{}^1 I_{1,1}{}^0 G_0{}^0 I_{0,2}{}^0 I_{0,1}{}^1\}$, tube $T_4 = \{G_1{}^0 I_{1,1}{}^1 G_0{}^0 I_{0,2}{}^1 I_{0,1}{}^0, G_1{}^0 I_{1,1}{}^1 G_0{}^0 I_{0,2}{}^0 I_{0,1}{}^0\}$. This is to say that the contents of the fifth tape square and the fourth tape square in each tape in the fourth **BMPDTM** are written by the corresponding read-write head and are, respectively, $G_1{}^1$ and $I_{1,1}{}^0$, and the contents of the fifth tape square and the fourth tape square in each tape in the fifth **BMPDTM** are written by the corresponding read-write head and are, respectively, $G_1{}^0$ and $I_{1,1}{}^1$. Simultaneously, the position of each read-write head is moved to the left new tape square, and the status of each finite state control is changed as "$G_1 = 1$" or "$G_1 = 0$".

Then, after the *first* execution of Step (9) is finished, tube $T_3 = \varnothing$, tube $T_4 = \varnothing$, tube $T_0 = \{G_1{}^1 I_{1,1}{}^0 G_0{}^0 I_{0,2}{}^1 I_{0,1}{}^1, G_1{}^1 I_{1,1}{}^0 G_0{}^0 I_{0,2}{}^0 I_{0,1}{}^1, G_1{}^1 I_{1,1}{}^1 G_0{}^0 I_{0,2}{}^1 I_{0,1}{}^0, G_1{}^0 I_{1,1}{}^1 G_0{}^0 I_{0,2}{}^0 I_{0,1}{}^0\}$. This implies that the execution environment for the two **BMDTMs** in the fourth **BMPDTM** and the two **BMDTMs** in the fifth **BMPDTM** becomes the first **BMPDTM**. Simultaneously, the position of each read-write head and the status of each finite state control are reserved.

Step (10) is the second loop and the lower bound and the upper bound are both two. After the first execution for Steps (10a), (10b) and (10c) is performed, tube $T_5 = \varnothing$, tube $T_6 = \varnothing$, tube $T_0 = \varnothing$, tube $T_7 = \{G_1^{\ 1}\ I_{1,\ 1}^{\ 0}\ G_0^{\ 0}\ I_{0,\ 2}^{\ 1}\ I_{0,\ 1}^{\ 1}\}$, tube $T_8 = \{G_1^{\ 0}\ I_{1,\ 1}^{\ 1}\ G_0^{\ 0}\ I_{0,\ 2}^{\ 1}\ I_{0,\ 1}^{\ 0}\}$, tube $T_9 = \{G_1^{\ 1}\ I_{1,\ 1}^{\ 0}\ G_0^{\ 0}\ I_{0,\ 2}^{\ 0}\ I_{0,\ 1}^{\ 1}\}$ and tube $T_{10} = \{G_1^{\ 0}\ I_{1,\ 1}^{\ 1}\ G_0^{\ 0}\ I_{0,\ 2}^{\ 0}\ I_{0,\ 1}^{\ 0}\}$. Next, after the first execution for Steps (10d), (10f), (10h) and (10j) is finished, each step returns a "yes". Thus, after the first execution for Steps (10e), (10g), (10i) and (10k) is implemented, tube $T_7 = \{G_2^{\ 1}\ I_{1,\ 2}^{\ 0}\ G_1^{\ 1}\ I_{1,\ 1}^{\ 0}\ G_0^{\ 0}\ I_{0,\ 2}^{\ 1}\ I_{0,\ 1}^{\ 1}\}$, tube $T_8 = \{G_2^{\ 0}\ I_{1,\ 2}^{\ 1}\ G_1^{\ 0}\ I_{1,\ 1}^{\ 1}\ G_0^{\ 0}\ I_{0,\ 2}^{\ 1}\ I_{0,\ 1}^{\ 0}\}$, tube $T_9 = \{G_2^{\ 0}\ I_{1,\ 2}^{\ 1}\ G_1^{\ 1}\ I_{1,\ 1}^{\ 0}\ G_0^{\ 0}\ I_{0,\ 2}^{\ 0}\ I_{0,\ 1}^{\ 1}\}$ and tube $T_{10} = \{G_2^{\ 0}\ I_{1,\ 2}^{\ 0}\ G_1^{\ 0}\ I_{1,\ 1}^{\ 1}\ G_0^{\ 0}\ I_{0,\ 2}^{\ 0}\ I_{0,\ 1}^{\ 0}\}$. Then, after the first execution for Step (10l) is performed, tube $T_0 = \{G_1^{\ 1}\ I_{1,\ 2}^{\ 0}\ G_1^{\ 1}\ I_{1,\ 1}^{\ 0}\ G_0^{\ 0}\ I_{0,\ 2}^{\ 1}\ I_{0,\ 1}^{\ 1},\ G_2^{\ 0}\ I_{1,\ 2}^{\ 1}\ G_1^{\ 0}\ I_{1,\ 1}^{\ 1}\ G_0^{\ 0}\ I_{0,\ 2}^{\ 1}\ I_{0,\ 1}^{\ 0},\ G_2^{\ 0}\ I_{1,\ 2}^{\ 1}\ G_1^{\ 1}\ I_{1,\ 1}^{\ 0}\ G_0^{\ 0}\ I_{0,\ 2}^{\ 0}\ I_{0,\ 1}^{\ 1},\ G_2^{\ 0}\ I_{1,\ 2}^{\ 0}\ G_1^{\ 0}\ I_{1,\ 1}^{\ 1}\ G_0^{\ 0}\ I_{0,\ 2}^{\ 0}\ I_{0,\ 1}^{\ 0}\}$ and other tubes become all empty tubes. Figure 7.46 is applied to show the result and **Algorithm 7.5** is terminated.

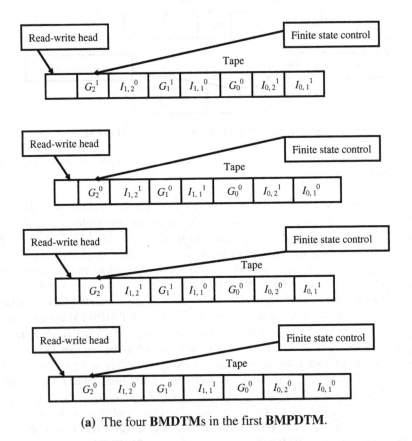

(a) The four **BMDTMs** in the first **BMPDTM**.

Fig. 7.46 Schematic representation of the current status of the execution environment to the first **BMPDTM**

7.5 Introduction to the Decrease Operation on Bio-molecular Computer

The decrease operation is employed to perform "$D = D - 1$", where D is unsigned integers of n bits. A symbol "$-$" is applied to represent the operation of decrease, and the expression, $D = D - 1$, can be rewritten as another expression: $D -$. An unsigned integer of n bits, D, is regarded as the minuend and the difference in the expression $D -$. The value "1" is also regarded as the subtrahend in the expression $D -$. Suppose that D can be represented as $D_{e, n}, D_{e, n - 1}, ..., D_{e, 1}$ for $0 \le e \le 1$, where the value for each $D_{e, k}$ for $1 \le k \le n$ is 1 or 0. Also assume that two binary numbers, H_k and $H_{k - 1}$ for $1 \le k \le n$, are employed to respectively represent the borrow for the minuend bit and the subtrahend bit and the previous borrow.

In the processing of computing $D -$, assume that the *original* value for D is represented as $D_{0, n}, D_{0, n - 1}, ..., D_{0, 1}$, and the difference to $D ++$ is represented as $D_{1, n}, D_{1, n - 1}, ..., D_{1, 1}$. Table 7.11 is regarded as the truth table of a one-bit subtractor for $D_{0, 1}, H_0, D_{1, 1}$ and H_1, and Table 7.12 is also regarded as the truth table of a one-bit subtractor for $D_{0, k}, H_{k - 1}, D_{1, k}$, and H_k for $2 \le k \le n$.

Table 7.11 The truth table of a one-bit subtractor for $D_{0, 1}, H_0, D_{1, 1}$ and H_1

$D_{0, 1}$	The first bit of the subtrahend	H_0	$D_{1, 1}$	H_1
0	1	0	1	1
1	1	0	0	0

Table 7.12 The truth table of a one-bit subtractor for $D_{0, k}, H_{k - 1}, D_{1, k}$, and H_k for $2 \le k \le n$

$D_{0, k}$	The kth bit of the subtrahend	$H_{k - 1}$	$D_{1, k}$	H_k
0	0	0	0	0
0	0	1	1	1
1	0	0	1	0
1	0	1	0	0

7.5.1 The Construction for the Parallel Operation of Decrease on Bio-molecular Computer

For the purpose of convenience, $D_{e, k}{}^1$ for $0 \le e \le 1$ and $1 \le k \le n$ denotes the fact that the value of $D_{e, k}$ is 1, $D_{e, k}{}^0$ for $0 \le e \le 1$ and $1 \le k \le n$ denotes the fact that the value of $D_{e, k}$ is 0, $H_k{}^1$ and $H_{k - 1}{}^1$ denote the fact that the value of H_k is one and the value of $H_{k - 1}$ is also one, and $H_k{}^0$ and $H_{k - 1}{}^0$ denote the fact that the value of H_k is zero and the value of $H_{k - 1}$ is also zero. The following algorithm is proposed to perform the parallel operation of decrease.

Algorithm 7.6: ParallelDecrease(T_0)

(1) Append-head($T_1, D_{0,1}{}^1$).

(2) Append-head($T_2, D_{0,1}{}^0$).

(3) $T_0 = \cup(T_1, T_2)$.

(4) **For** $k = 2$ **to** n

      (4a) Amplify(T_0, T_1, T_2).

      (4b) Append-head($T_1, D_{0,k}{}^1$).

      (4c) Append-head($T_2, D_{0,k}{}^0$).

      (4d) $T_0 = \cup(T_1, T_2)$.

EndFor

(5) Append-head($T_0, H_0{}^0$).

(6) $T_3 = +(T_0, D_{0,1}{}^1)$ and $T_4 = -(T_0, D_{0,1}{}^1)$.

(7) **If** (Detect(T_3) $= =$ "yes") **Then**

      (7a) Append-head($T_3, D_{1,1}{}^0$) and Append-head($T_3, H_1{}^0$).

    EndIf

(8) **If** (Detect(T_4) $= =$ "yes") **Then**

      (8a) Append-head($T_4, D_{1,1}{}^1$) and Append-head($T_4, H_1{}^1$).

    EndIf

(9) $T_0 = \cup(T_3, T_4)$.

(10) **For** $k = 2$ **to** n

(10a) $T_5 = +(T_0, D_{0,k}{}^1)$ and $T_6 = -(T_0, D_{0,k}{}^1)$.

(10b) $T_7 = +(T_5, H_{k-1}{}^1)$ and $T_8 = -(T_5, H_{k-1}{}^1)$.

(10c) $T_9 = +(T_6, H_{k-1}{}^1)$ and $T_{10} = -(T_6, H_{k-1}{}^1)$.

(10d) **If** (Detect(T_7) $= =$ "yes") **Then**

      (10e) Append-head($T_7, D_{1,k}{}^0$) and Append-head($T_7, H_k{}^0$).

    EndIf

(10f) **If** (Detect(T_8) $= =$ "yes") **Then**

      (10g) Append-head($T_8, D_{1,k}{}^1$) and Append-head($T_8, H_k{}^0$).

    EndIf

(10h) **If** (Detect(T_9) $= =$ "yes") **Then**

      (10i) Append-head($T_9, D_{1,k}{}^1$) and Append-head($T_9, H_k{}^1$).

    EndIf

(10j) **If** (Detect(T_{10}) $= =$ "yes") **Then**

      (10k) Append-head($T_{10}, D_{1,k}{}^0$) and Append-head($T_{10}, H_k{}^0$).

    EndIf

    (10l) $T_0 = \cup(T_7, T_8, T_9, T_{10})$.

EndFor

EndAlgorithm

Lemma 7-6: *The algorithm, **ParallelDecrease(T_0)**, can be applied to carry out the parallel operation of decrease.*

Proof The algorithm, **ParallelDecrease**(T_0), is implemented by means of the *extract, amplify, detect, append-head* and *merge* operations. Solution space of 2^n unsigned integers of n bits (the range of values for them is from 0 to $2^n - 1$) is constructed from each execution for Steps (1) through (4d). After those operations are finished, 2^n combinations of n bits are included in tube T_0. Since the previous borrow for subtraction of the first bit to each minuend and subtrahend is zero, from the execution of Step (5), the value "0" of H_0 is appended into the head of each bit pattern in tube T_0.

From each execution of Step (6), tube T_3 inludes all of the inputs that have $D_{0, 1} = 1$, representing the first column of the second row in Table 7.11, and tube T_4 contains of all of the inputs that have $D_{0, 1} = 0$, representing the first column of the first row in Table 7.11. Because the contents for tubes T_3 and T_4 are not empty, a "yes" is returned from Step (7) and Step (8). Hence, from each execution of Step (7a), the value "0" of $D_{1, 1}$ and the value "0" of H_1 are appended into the head of each bit pattern in tube T_3, and from each execution of Step (8a), the value "1" of $D_{1, 1}$ and the value "1" of H_1 are appended into the head of each bit pattern in tube T_4. This indicates that Table 7.11 is finished. Next, each execution of Step (9) uses the *merge* operation to pour tubes T_3 through T_4 into tube T_0. Tube T_0 consists of the result performing Table 7.11.

Step (10) is the second loop and is employed to carry out the parallel one-bit subtractor of $(n - 1)$ times. From each execution for Steps (10a) through (10c), T_5 contains all of the inputs that have $D_{0, k} = 1$, T_6 includes all of the inputs that have $D_{0, k} = 0$, T_7 consists of all of the inputs that have $D_{0, k} = 1$ and $H_k = 1$, T_8 contains all of the inputs that have $D_{0, k} = 1$ and $H_k = 0$, T_9 includes all of the inputs that have $D_{0, k} = 0$ and $H_k = 1$, and T_{10} includes all of the inputs that have $D_{0, k} = 0$ and $H_k = 0$. Having performed Steps (10a) through (10c), this indicates that four different inputs of a one-bit subtractor as shown in Table 7.12 were poured into tubes T_7 through T_{10}, respectively.

From each execution for Steps (10d), (10f), (10h) and (10j), since the contents for tubes T_7, T_8, T_9 and T_{10} are all not empty, therefore, a "yes" is returned from each step. Next, on each execution for Steps (10e), (10g), (10i) and (10k), the *append-head* operations are applied to append $D_{1, k}^{1}$ or $D_{1, k}^{0}$, and H_k^{1} or H_k^{0} onto the head of every bit pattern in the corresponding tubes. After finishing Steps (10a) through (10k), this implies that four different outputs of a one-bit subtractor in Table 7.12 are appended into tubes T_7 through T_{10}. Next, each execution of Step (10l) uses the *merge* operation to pour tubes T_7 through T_{10} into tube T_0. Tube T_0 contains the result performing Table 7.12. Repeat execution of Steps (10a) through (10l) until the most significant bit for the minuend and the subtrahend is processed. Tube T_0 obtains the result carrying out the parallel operation of decrease for 2^n unsigned integers of n bits.   ∎

7.5.2 The Power for the Parallel Operation of Decrease on Bio-molecular Computer

Consider that four values for an unsigned integer of two bits are, subsequently, $00(0_{10})$ ($D_{0,\,2}{}^0 D_{0,\,1}{}^0$), $01(1_{10})$ ($D_{0,\,2}{}^0 D_{0,\,1}{}^1$), $10(2_{10})$ ($D_{0,\,2}{}^1 D_{0,\,1}{}^0$), and $11(3_{10})$ ($D_{0,\,2}{}^1 D_{0,\,1}{}^1$). We want to simultaneously decrease each value of the four values. **Algorithm 7.6**, **ParallelDecrease**(T_0), can be applied to finish the computational task. Tube T_0 is an empty tube and is regarded as an input tube of **Algorithm 7.6**. From Definition 5−2, the input tube T_0 can be regarded as the execution environment of the first **BMPDTM**. Similarly, each tube T_k in **Algorithm 7.6** for $1 \leq k \leq 10$ can be also regarded as the execution environment of the $(k+1)$th **BMPDTM**.

A **BMPDTM** with four bio-molecular deterministic one-tape Turing machines from Steps (1) through (4d) in **Algorithm 7.6** is constructed. After the first execution for Step (1) and Step (2) is implemented, tube $T_1 = \{D_{0,\,1}{}^1\}$ and tube $T_2 = \{D_{0,\,1}{}^0\}$. This is to say that a **BMDTM** in the second **BMPDTM** and in the third **BMPDTM** is produced. Figure 7.47 is used to show the current status of the execution environment to the second **BMPDTM** and the third **BMPDTM**. From Fig. 7.47, the content of the first tape square for the tape in the first **BMDTM** in the second **BMPDTM** is written by its corresponding read-write head and is 1 ($D_{0,\,1} = 1$), and the content of the first tape square for the tape in the first **BMDTM** in the third **BMPDTM** is written by its corresponding read-write head and is 0 ($D_{0,\,1} = 0$). Simultaneously, for the two **BMDTMs**, the position of the corresponding read-write head is moved to the *left new* tape square, and the status of the corresponding finite state control is, respectively, "$D_{0,\,1} = 1$" and "$D_{0,\,1} = 0$".

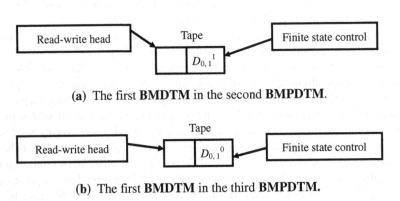

(a) The first **BMDTM** in the second **BMPDTM**.

(b) The first **BMDTM** in the third **BMPDTM**.

Fig. 7.47 Schematic representation of the current status of the execution environment to the second **BMPDTM** and the third **BMPDTM**

Next, after the execution for Step (3) is finished, tube $T_0 = \{D_{0,1}{}^1, D_{0,1}{}^0\}$, tube $T_1 = \emptyset$ and tube $T_2 = \emptyset$. This implies that the execution environment for the first **BMDTM** in the second **BMPDTM** and the first **BMDTM** in the third **BMPDTM** becomes the first **BMPDTM**. From Fig. 7.48, the contents to the two tapes in the execution environment of the first **BMPDTM** are not changed, and the position of each read-write head and the status of each finite state control are reserved.

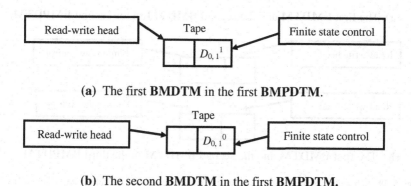

(a) The first **BMDTM** in the first **BMPDTM**.

(b) The second **BMDTM** in the first **BMPDTM**.

Fig. 7.48 Schematic representation of the current status of the execution environment to the first **BMPDTM**

Step (4) is the first loop and the upper bound (n) is two since the number of bits for representing those four values is two. Thus, after the first execution of Step (4a) is performed, tube $T_0 = \emptyset$, tube $T_1 = \{D_{0,1}{}^1, D_{0,1}{}^0\}$ and tube $T_2 = \{D_{0,1}{}^1, D_{0,1}{}^0\}$. This indicates that the first **BMDTM** and the second **BMDTM** in the execution environment of the first **BMPDTM** are both copied into the second **BMPDTM** and the third **BMPDTM**. Figure 7.49 is employed to illustrate the current status of the execution environment to the second **BMPDTM** and the third **BMPDTM**. From Fig. 7.49, the contents of the first tape square for the corresponding tape of the first **BMDTM** and the corresponding tape of the second **BMDTM** in the execution environment of the second **BMPDTM** are, respectively, 1 ($D_{0,1} = 1$) and 0 ($D_{0,1} = 0$). The contents of the first tape square for the corresponding tape of the first **BMDTM** and the corresponding tape of the second **BMDTM** in the execution environment of the third **BMPDTM** are also, respectively, 0 ($D_{0,1} = 0$) and 1 ($D_{0,1} = 1$). From Fig. 7.49, it is pointed out that four bio-molecular deterministic one-tape Turing machines are constructed.

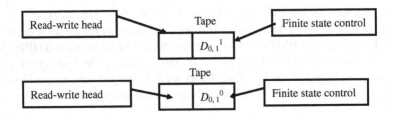

(a) The first **BMDTM** and the second **BMDTM** in the second **BMPDTM**.

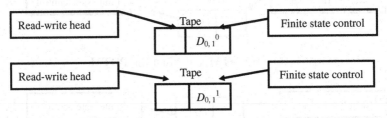

(b) The first **BMDTM** and the second **BMDTM** in the third **BMPDTM**.

Fig. 7.49 Schematic representation of the current status of the execution environment to the second **BMPDTM** and the third **BMPDTM**

Next, after the first execution for Step (4b) and Step (4c) is finished, tube $T_1 = \{D_{0,2}{}^1 D_{0,1}{}^1, D_{0,2}{}^1 D_{0,1}{}^0\}$ and tube $T_2 = \{D_{0,2}{}^0 D_{0,1}{}^1, D_{0,2}{}^0 D_{0,1}{}^0\}$. This implies that the content of the *second* tape square for the tape in the first **BMDTM** in the second **BMPDTM** is written by its corresponding read-write head and is 1 ($D_{0,2} = 1$), and the content of the *second* tape square for the tape in the second **BMDTM** in the second **BMPDTM** is written by its corresponding read-write head and is also 1 ($D_{0,2} = 1$). Similarly, the content of the *second* tape square for the tape in the first **BMDTM** in the third **BMPDTM** is written by its corresponding read-write head and is 0 ($D_{0,2} = 0$), and the content of the *second* tape square for the tape in the second **BMDTM** in the third **BMPDTM** is written by its corresponding read-write head and is also 0 ($D_{0,2} = 0$). Figure 7.50 is applied to reveal the current status of the execution environment to the second **BMPDTM** and the third **BMPDTM**. From Fig. 7.50, the position of each read-write head is moved to the left new tape square, and the status of each finite state control is changed as "$D_{0,2} = 1$" and "$D_{0,2} = 0$".

Next, after the first execution for Step (4d) is performed, tube $T_0 = \{D_{0,2}{}^1 D_{0,}$ $_1{}^1, D_{0,2}{}^1 D_{0,1}{}^0, D_{0,2}{}^0 D_{0,1}{}^1, D_{0,2}{}^0 D_{0,1}{}^0\}$, tube $T_1 = \varnothing$ and tube $T_2 = \varnothing$. This is to say that the execution environment for those bio-molecular deterministic one-tape Turing machines in the second **BMPDTM** and in the third **BMPDTM**

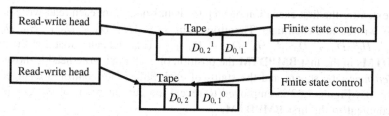

(a) The first **BMDTM** and the second **BMDTM** in the second **BMPDTM**.

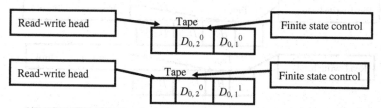

(b) The first **BMDTM** and the second **BMDTM** in the third **BMPDTM**.

Fig. 7.50 Schematic representation of the current status of the execution environment to the second **BMPDTM** and the third **BMPDTM**

becomes the first **BM-PDTM**. Figure 7.51 is used to show the current status of the execution environment to the first **BMPDTM**. From Fig. 7.51, the contents to the four tapes in the execution environment of the first **BMPDTM** are not changed, and the position of the corresponding read-write head and the state of the corresponding finite state control are reserved.

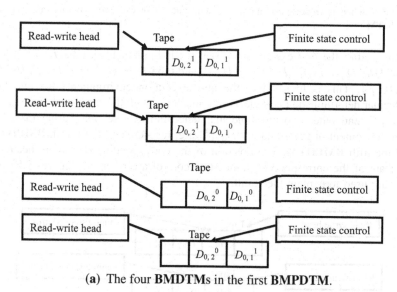

(a) The four **BMDTMs** in the first **BMPDTM**.

Fig. 7.51 Schematic representation of the current status of the execution environment to the first **BMPDTM**

Next, after the first execution of Step (5) is finished, from each read-write head, $H_0{}^0$ is written into each tape. Therefore, tube $T_0 = \{H_0{}^0 D_{0,\,2}{}^1 D_{0,\,1}{}^1, H_0{}^0 D_{0,\,2}{}^1 D_{0,\,1}{}^0, H_0{}^0 D_{0,\,2}{}^0 D_{0,\,1}{}^1, H_0{}^0 D_{0,\,2}{}^0 D_{0,\,1}{}^0\}$. This indicates that for the four **BMDTM**s in the first **BMPDTM** the position of each read-write head is moved to the *left new* tape square and the status of each finite state control is changed as "$H_0 = 0$". Figure 7.52 is employed to reveal the current status of the execution environment to the first **BMPDTM**.

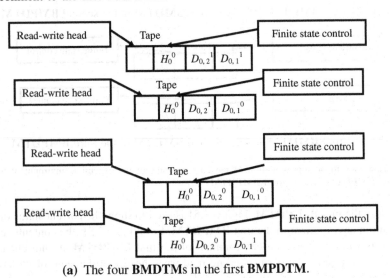

(a) The four **BMDTM**s in the first **BMPDTM**.

Fig. 7.52 Schematic representation of the current status of the execution environment to the first **BMPDTM**

Next, after the *first* execution of Step (6) is implemented, tube $T_0 = \emptyset$, tube $T_3 = \{H_0{}^0 D_{0,\,2}{}^1 D_{0,\,1}{}^1, H_0{}^0 D_{0,\,2}{}^0 D_{0,\,1}{}^1\}$ and tube $T_4 = \{H_0{}^0 D_{,\,2}{}^1 D_{0,\,1}{}^0, H_0{}^0 D_{0,\,2}{}^0 D_{0,\,1}{}^0\}$. This is to say that the new execution environments for two bio-molecular deterministic one-tape Turing machines with the content of tape square, "$D_{0,\,1}{}^1$", and other two bio-molecular deterministic one-tape Turing machines with the content of tape square, "$D_{0,\,1}{}^0$" are, respectively, the fourth **BMPDTM** and the fifth **BMPDTM**. The position of the corresponding read-write head and the state of the corresponding finite state control are reserved. Figures 7.53 and 7.54 are used to explain the result.

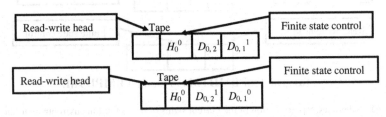

(a) The two **BMDTM**s in the fourth **BMPDTM**.

Fig. 7.53 Schematic representation of the current status of the execution environment to the fourth **BMPDTM**

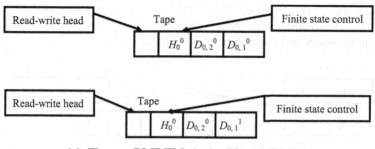

(a) The two **BMDTM**s in the fifth BMPDTM.

Fig. 7.54 Schematic representation of the current status of the execution environment to the fifth **BMPDTM**

Because tube $T_3 \neq \emptyset$ and tube $T_4 \neq \emptyset$, from the first execution of Step (7) and the first execution of Step (8), the conditions are true. Hence, after the first execution for Steps (7a) and (8a) is finished, tube $T_3 = \{H_1{}^0 D_{1,\,1}{}^0 H_0{}^0 D_{0,\,2}{}^0 D_{0,\,1}{}^1, H_1{}^0 D_{1,\,1}{}^0 H_0{}^0 D_{0,\,2}{}^0 D_{0,\,1}{}^1\}$, tube $T_4 = \{H_1{}^1 D_{1,\,1}{}^1 H_0{}^0 D_{0,\,2}{}^0 D_{0,\,1}{}^0, H_1{}^1 D_{1,\,1}{}^1 H_0{}^0 D_{0,\,2}{}^0 D_{0,\,1}{}^0\}$. This implies that the contents of the fifth tape square and the fourth tape square in each tape in the fourth **BMPDTM** are written by the corresponding read-write head and are, respectively, $H_1{}^0$ and $D_{1,\,1}{}^0$, and the contents of the fifth tape square and the fourth tape square in each tape in the fifth **BMPDTM** are written by the corresponding read-write head and are, respectively, $H_1{}^1$ and $D_{1,\,1}{}^1$. Simultaneously, the position of each read-write head is moved to the left new tape square, and the status of each finite state control is changed as "$H_1 = 1$" or "$H_1 = 0$".

Then, after the *first* execution of Step (9) is finished, tube $T_3 = \emptyset$, tube $T_4 = \emptyset$, tube $T_0 = \{H_1{}^0 D_{1,\,1}{}^0 H_0{}^0 D_{0,\,2}{}^1 D_{0,\,1}{}^1, H_1{}^0 D_{1,\,1}{}^0 H_0{}^0 D_{0,\,2}{}^0 D_{0,\,1}{}^1, H_1{}^1 D_{1,\,1}{}^1 H_0{}^0 D_{0,\,2}{}^1 D_{0,\,1}{}^0, H_1{}^1 D_{1,\,1}{}^1 H_0{}^0 D_{0,\,2}{}^0 D_{0,\,1}{}^0\}$. This is to say that the execution environment for the two **BMDTM**s in the fourth **BMPDTM** and the two **BMDTM**s in the fifth **BMPDTM** becomes the first **BMPDTM**. Simultaneously, the position of each read-write head and the status of each finite state control are reserved.

Step (10) is the second loop and the lower bound and the upper bound are both two. After the first execution for Steps (10a), (10b) and (10c) is finished, tube $T_5 = \emptyset$, tube $T_6 = \emptyset$, tube $T_0 = \emptyset$, tube $T_7 = \{H_1{}^1 D_{1,\,1}{}^1 H_0{}^0 D_{0,\,2}{}^1 D_{0,\,1}{}^0\}$, tube $T_8 = \{H_1{}^0 D_{1,\,1}{}^0 H_0{}^0 D_{0,\,2}{}^0 D_{0,\,1}{}^1\}$, tube $T_9 = \{H_1{}^1 D_{1,\,1}{}^1 H_0{}^0 D_{0,\,2}{}^0 D_{0,\,1}{}^0\}$ and tube $T_{10} = \{H_1{}^0 D_{1,\,1}{}^0 H_0{}^0 D_{0,\,2}{}^0 D_{0,\,1}{}^1\}$. Next, after the first execution for Steps (10d), (10f), (10h) and (10j) is finished, each step returns a "yes". Thus, after the first execution for Steps (10e), (10g), (10i) and (10k) is performed, tube $T_7 = \{H_2{}^0 D_{1,\,2}{}^0 H_1{}^1 D_{1,\,1}{}^1 H_0{}^0 D_{0,\,2}{}^1 D_{0,\,1}{}^0\}$, tube $T_8 = \{H_2{}^0 D_{1,\,2}{}^0 H_1{}^0 D_{1,\,1}{}^0 H_0{}^0 D_{0,\,2}{}^0 D_{0,\,1}{}^1\}$, tube $T_9 = \{H_2{}^1 D_{1,\,2}{}^1 H_1{}^1 D_{1,\,1}{}^1 H_0{}^0 D_{0,\,2}{}^0 D_{0,\,1}{}^0\}$ and tube $T_{10} = \{H_2{}^0 D_{1,\,2}{}^0 H_1{}^0 D_{1,\,1}{}^0 H_0{}^0 D_{0,\,2}{}^0 D_{0,\,1}{}^1\}$. Then, after the first execution for Step (10l) is performed, tube $T_0 = \{H_2{}^0 D_{1,\,2}{}^0 H_1{}^1 D_{1,\,1}{}^1 H_0{}^0 D_{0,\,2}{}^1 D_{0,\,1}{}^0, H_2{}^0 D_{1,\,2}{}^1 H_1{}^0 D_{1,\,1}{}^0$

$H_0{}^0 D_{0,2}{}^1 D_{0,1}{}^1, H_2{}^1 D_{1,2}{}^1 H_1{}^1 D_{1,1}{}^1 H_0{}^0 D_{0,2}{}^0 D_{0,1}{}^0, H_2{}^0 D_{1,2}{}^0 H_1{}^0 D_{1,1}{}^0 H_0{}^0$ $D_{0,2}{}^0 D_{0,1}{}^1\}$ and other tubes become all empty tubes. Figure 7.55 is employed to illustrate the result and Algorithm 7.6 is terminated.

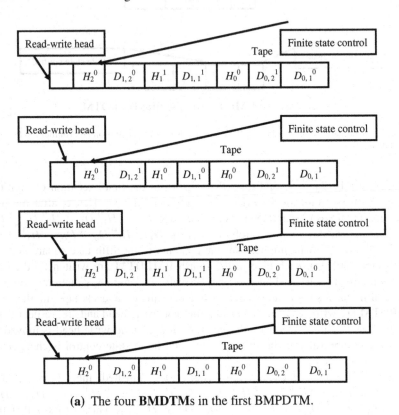

(a) The four **BMDTMs** in the first BMPDTM.

Fig. 7.55 Schematic representation of the current status of the execution environment to the first **BMPDTM**

7.6 Introduction for Finding the Maximum and Minimum Numbers of One on Bio-molecular Computer

Consider that four combinations of two bits that are, subsequently, $00(0_{10})$, $01(1_{10})$, $10(2_{10})$, and $11(3_{10})$. One interesting question is how the four combinations are classified from the number of one in their combinations. Because the numbers of one for $11(3_{10})$, $10(2_{10})$, $01(1_{10})$ and $00(0_{10})$ are, subsequently, two, one, one and zero, $11(3_{10})$ and $00(0_{10})$ are two different classification and $10(2_{10})$ and $01(1_{10})$ are the same classification. Similarly, we can extend the interesting question that is how the 2^n combinations of n bits are classified from the number of one in their combinations. This is to say that those combinations have k ones for $0 \leq k \leq n$.

7.6.1 The Construction for Finding the Maximum and Minimum Numbers of One on Bio-molecular Computer

Assume that a binary number of n bits, $P_n, P_{n-1}, \ldots, P_2, P_1$ can be applied to form 2^n combinations, where the value for each P_k is one or zero for $1 \leq k \leq n$. For the sake of convenience, P_k^1 for $1 \leq k \leq n$ denotes the fact that the value of P_k is one and P_k^0 denotes the fact that the value of P_k is zero for $1 \leq k \leq n$. The following algorithm is proposed to find the maximum and minimum numbers of one from 2^n combinations.

Algorithm 7.7: ParallelFind(T_0)

(1) Append-head(T_1, P_1^1).

(2) Append-head(T_2, P_1^0).

(3) $T_0 = \cup(T_1, T_2)$.

(4) **For** $k = 2$ **to** n

    (4a) Amplify(T_0, T_1, T_2).

    (4b) Append-head(T_1, P_k^1).

    (4c) Append-head(T_2, P_k^0).

    (4d) $T_0 = \cup(T_1, T_2)$.

EndFor

(5) **For** $k = 0$ **to** $n - 1$

    (6) **For** $j = k$ **downto** 0

        (6a) $T_{j+1}^{ON} = +(T_j, P_{k+1}^1)$ and $T_j = -(T_j, P_{k+1}^1)$.

        (6b) $T_{j+1} = \cup(T_{j+1}, T_{j+1}^{ON})$.

    EndFor

EndFor

EndAlgorithm

Lemma 7-7: *The algorithm, **ParallelFind**(T_0), can be used to find the maximum and minimum numbers of one from 2^n combinations of n bits.*

Proof The algorithm, **ParallelFind**(T_0), is implemented by means of the *extract, amplify, append-head* and *merge* operations. Solution space of 2^n states of n bits is generated from each execution for Steps (1) through (4d). After those operations are performed, 2^n combinations of n bits are contained in tube T_0.

Step (5) and Step (6) are, respectively, the outer loop and the inner loop of the nested loop. Because the loop index variable k is from 0 to $n - 1$, Step (5) and Step (6) are mainly employed to figure out the influence of $P_{k + 1}$ for the number of one in tubes T_0 through $T_{j + 1}$ for that the value of j is from k through 0. On each execution of Step (6a), it uses the *extract* operation from tube T_j to form two different tubes: $T_{j + 1}{}^{ON}$ and T_j. This implies that tube $T_{j + 1}{}^{ON}$ contains those combinations that have $P_{k + 1} = 1$ and tube T_j includes those combinations that have $P_{k + 1} = 0$. Because those combinations in tube T_j have j ones, those combinations in $T_{j + 1}{}^{ON}$ have $(j + 1)$ ones. Next, each execution of Step (6b) applies the *merge* operation to pour tube $T_{j + 1}{}^{ON}$ into tube $T_{j + 1}$. This is to say that those combinations in tube $T_{j + 1}$ have $(j + 1)$ ones. Repeat to execute (6a) and (6b) until the influence of P_n for the number of one in tubes T_0 through T_n is processed. This implies that those combinations in tube T_k for $0 \le k \le n$ have k ones.   ∎

7.6.2 The Power for Finding the Maximum and Minimum Numbers of One on Bio-molecular Computer

Consider that four states of two bits are, respectively, $00(0_{10})$ $(P_2{}^0 P_1{}^0)$, $01(1_{10})$ $(P_2{}^0 P_1{}^1)$, $10(2_{10})$ $(P_2{}^1 P_1{}^0)$, and $11(3_{10})$ $(P_2{}^1 P_1{}^1)$. We want to find the maximum number of one and the minimum number of one from the four states of two bits. **Algorithm 7.7, ParallelFind**(T_0), can be employed to carry out the searching task. Tube T_0 is an empty tube and is regarded as an input tube of **Algorithm 7.7**. From Definition 5–2, the input tube T_0 and other tubes can be regarded as different **BMPDTM**s.

A **BMPDTM** with four bio-molecular deterministic one-tape Turing machines from Steps (1) through (4d) in **Algorithm 7.7** is generated. After the first execution for Step (1) and Step (2) is performed, tube $T_1 = \{P_1{}^1\}$ and tube $T_2 = \{P_1{}^0\}$. This implies that a **BMDTM** in tube T_1 and in tube T_2 is constructed. Figure 7.56 is applied to illustrate the result. From Fig. 7.56, the content of the first tape square for the tape in the first **BMDTM** in tube T_1 is written by its corresponding read-write head and is 1 ($P_1 = 1$), and the content of the first tape square for the tape in the first **BMDTM** in tube T_2 is written by its corresponding read-write head and is 0 ($P_1 = 0$). Simultaneously, for the two **BMDTM**s, the position of the corresponding read-write head is moved to the *left new* tape square, and the status of the corresponding finite state control is, respectively, "$P_1 = 1$" and "$P_1 = 0$".

Next, after the execution for Step (3) is implemented, tube $T_0 = \{P_1^1, P_1^0\}$, tube $T_1 = \varnothing$ and tube $T_2 = \varnothing$. This is to say that the execution environment for the first **BMDTM** in tube T_1 and the first **BMDTM** in tube T_2 becomes tube T_0. From Fig. 7.57, the contents to the two tapes in tube T_0 are not changed, and the position of each read-write head and the status of each finite state control are reserved.

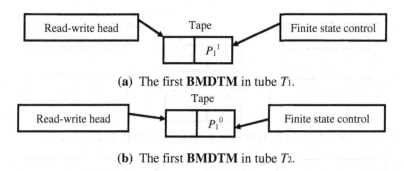

(a) The first **BMDTM** in tube T_1.

(b) The first **BMDTM** in tube T_2.

Fig. 7.56 Schematic representation of the current status of the execution environment to tube T_1 and tube T_2

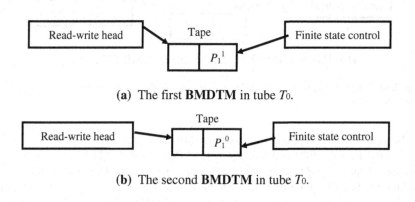

(a) The first **BMDTM** in tube T_0.

(b) The second **BMDTM** in tube T_0.

Fig. 7.57 Schematic representation of the current status of the execution environment to tube T_0

Step (4) is the first loop and the upper bound (n) is two since the number of bits for representing those four combinations is two. Therefore, after the first execution of Step (4a) is implemented, tube $T_0 = \varnothing$, tube $T_1 = \{P_1^1, P_1^0\}$ and tube $T_2 = \{P_1^1, P_1^0\}$. This implies that the first **BMDTM** and the second **BMDTM** in tube T_0 are both copied into tube T_1 and tube T_2. Figure 7.58 is used to reveal the result. From Fig. 7.58, the contents of the first tape square for the corresponding tape of the first **BMDTM** and the corresponding tape of the second **BMDTM** in tube T_1 are, respectively, 1 ($P_1 = 1$) and 0 ($P_1 = 0$). The contents of the first tape square for the corresponding tape of the first **BMDTM** and the corresponding tape of the second **BMDTM** in tube T_2 are also, respectively, 0 ($P_1 = 0$) and 1 ($P_1 = 1$). From Fig. 7.58, four bio-molecular deterministic one-tape Turing machines are constructed and the position of each read-write head and the status of each finite state control are reserved.

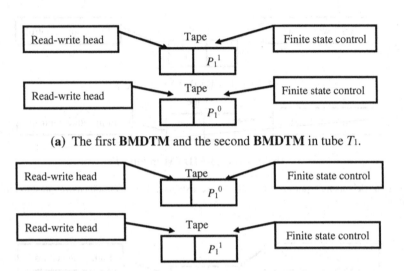

(a) The first **BMDTM** and the second **BMDTM** in tube T_1.

(b) The first **BMDTM** and the second **BMDTM** in tube T_2.

Fig. 7.58 Schematic representation of the current status of the execution environment to tube T_1 and tube T_2

Next, after the first execution for Step (4b) and Step (4c) is performed, tube $T_1 = \{P_2{}^1 P_1{}^1, P_2{}^1 P_1{}^0\}$ and tube $T_2 = \{P_2{}^0 P_1{}^1, P_2{}^0 P_1{}^0\}$. This is to say that the content of the *second* tape square for the tape in the first **BMDTM** in tube T_1 is written by its corresponding read-write head and is 1 ($P_2 = 1$), and the content of the *second* tape square for the tape in the second **BMDTM** in tube T_1 is written by its corresponding read-write head and is also 1 ($P_2 = 1$). Similarly, the content of the *second* tape square for the tape in the first **BMDTM** in tube T_2 is written by its corresponding read-write head and is 0 ($P_2 = 0$), and the content of the *second* tape square for the tape in the second **BMDTM** in tube T_2 is written by its corresponding read-write head and is also 0 ($P_2 = 0$). From Fig. 7.59 is employed to explain the result, the position of each read-write head is moved to the left new tape square and the status of each finite state control is changed as "$P_2 = 1$" and "$P_2 = 0$".

Next, after the first execution for Step (4d) is implemented, tube $T_0 = \{P_2{}^1 P_1{}^1, P_2{}^1 P_1{}^0, P_2{}^0 P_1{}^1, P_2{}^0 P_1{}^0\}$, tube $T_1 = \varnothing$ and tube $T_2 = \varnothing$. This indicates that the

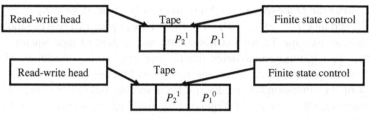

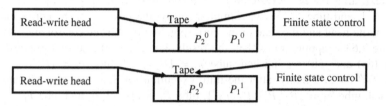

(a) The first **BMDTM** and the second **BMDTM** in tube T_1.

(b) The first **BMDTM** and the second **BMDTM** in tube T_2.

Fig. 7.59 Schematic representation of the current status of the execution environment to tube T_1 and tube T_2

execution environment for those bio-molecular deterministic one-tape Turing machines in tube T_1 and tube T_2 becomes tube T_0. Figure 7.60 is applied to show the result. From Fig. 7.60, the contents to the four tapes in tube T_0 are not changed, and the position of the corresponding read-write head and the state of the corresponding finite state control are reserved.

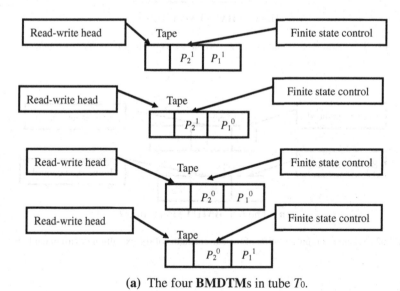

(a) The four **BMDTMs** in tube T_0.

Fig. 7.60 Schematic representation of the current status of the execution environment to tube T_0

Next, after the *first* execution of Step (6a) is implemented, tube $T_0 = \{P_2^1 P_1^0, P_2^0 P_1^0\}$, and tube $T_1^{ON} = \{P_2^1 P_1^1, P_2^0 P_1^1\}$. This implies that two bio-molecular deterministic one-tape Turing machines with the content of tape square, "P_1^1", and other two bio-molecular deterministic one-tape Turing machines with the content of tape square, "P_1^0" are, respectively, put into tube T_0 and tube T_1^{ON}. The position of the corresponding read-write head and the state of the corresponding finite state control are reserved. Figures 7.61 and 7.62 are employed to illustrate the result.

Next, after the *first* execution of Step (6b) is performed, tube $T_1 = \{P_2^1 P_1^1, P_2^0 P_1^1\}$, and tube $T_1^{ON} = \emptyset$. Simultaneously, the position of the corresponding read-write head and the state of the corresponding finite state control are reserved. Figure 7.63 is applied to explain the result. Then, after the second execution for Step (6a) and (6b) are finished, tube $T_2 = \{P_2^1 P_1^1\}$, tube $T_1 = \{P_2^0 P_1^1\}$, and tube $T_2^{ON} = \emptyset$. Finally, after the third execution for Step (6a) and (6b) are performed, tube $T_1 = \{P_2^0 P_1^1, P_2^1 P_1^0\}$, tube $T_0 = \{P_2^0 P_1^0\}$, and tube $T_1^{ON} = \emptyset$. Figure 7.64 is used to reveal the final result and **Algorithm 7.7** is terminated.

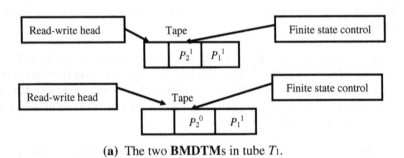

(a) The two **BMDTMs** in tube T_1.

Fig. 7.61 Schematic representation of the current status of the execution environment to tube T_1^{ON}

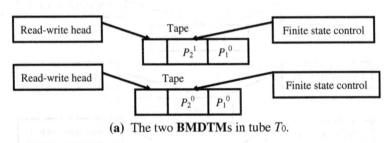

(a) The two **BMDTMs** in tube T_0.

Fig. 7.62 Schematic representation of the current status of the execution environment to tube T_0

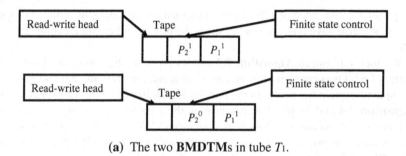

(a) The two **BMDTMs** in tube T_1.

Fig. 7.63 Schematic representation of the current status of the execution environment to tube T_1

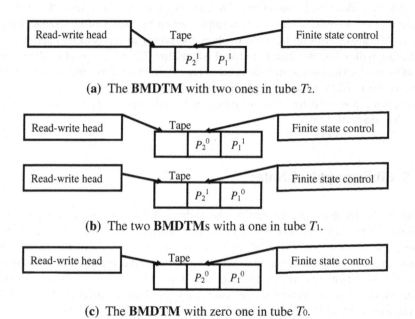

(a) The **BMDTM** with two ones in tube T_2.

(b) The two **BMDTMs** with a one in tube T_1.

(c) The **BMDTM** with zero one in tube T_0.

Fig. 7.64 Schematic representation of the current status of the execution environment to tube T_2, tube T_1 and tube T_0

7.7 Summary

In this chapter we provided an introduction to how comparators, shifters, increase, and decrease were constructed by means of molecular operations, and we also provided a description to how two specific operations which are to find the maximum number of "1" and to find the minimum number of "1" were implemented by means of biological operations. We illustrated **Algorithm 7.1** and **Algorithm 7.2** and their proof to reveal how the parallel comparator of a bit and the parallel comparator of n bit were implemented by means of biological operations. We also gave one example to show the power to the parallel comparator of n bit.

We then introduced **Algorithm 7.3** and its proof to show how the parallel left shifter of n bit was completed by means of molecular operations. We also gave one example to explain the power to the parallel left shifter of n bit. We then described **Algorithm 7.4** and its proof to reveal how the parallel right shifter of n bit was constructed by means of biological operations. We also gave one example to demonstrate the power to the parallel right shifter of n bit. We then illustrated **Algorithm 7.5** and its proof to explain how the parallel operation of increase with n bits was completed by means of biological operations. We also gave one example to explain the power to the parallel operation of increase with n bit.

We then described **Algorithm 7.6** and its proof to show how the parallel operation of decrease with n bit was implemented by means of molecular operations. We also gave one example to show the power to the parallel operation of decrease with n bit. We then introduced **Algorithm 7.7** and its proof to show how finding the maximum and minimum numbers of one from 2^n combinations of n bit was constructed by means of biological operations. We also gave one example to explain the power to find the maximum and minimum numbers of one from 2^n combinations of n bit.

7.8 Bibliographical Notes

The textbooks that were written by the authors in Brown and Vranesic (2007), Mano (1979), Mano (1993), Null and Lobur (2010), Shiva (2008), Stalling (2000) are good introduction for digital circuits of comparators, shifters, adders and subtractors. A good introduction to automata theory and complexity of problems is the textbook that was written by Hopcroft et al. [Hopcroft et al. 2006]. A good illustration to basic theorems and properties of Boolean algebra discussed in exercises in Sect. 7.9 is Brown and Vranesic (2007), Mano (1979).

7.9 Exercises

7.1 The binary operator $\vee$ defines logical operation **OR**. The truth table to a logical operation, $x \vee 1$, is shown in Table 7.13, where x is a Boolean variable that is the first input and 1 is the second input. Based on Table 7.13, write a bio-molecular program to show $x \vee 1 = 1$.

Table 7.13 The truth table to a logical operation, $x \vee 1$, is shown

The first input (x)	The second input	$x \vee 1$
0	1	1
1	1	1

7.2 The binary operator $\wedge$ defines logical operation **AND**. The truth table to a logical operation, $x \wedge 0$, is shown in Table 7.14, where x is a Boolean variable that is the first input and 0 is the second input. Based on Table 7.14, write a bio-molecular program to demonstrate $x \wedge 0 = 0$.

Table 7.14 The truth table to a logical operation, $x \wedge 0$, is shown	The first input (x)	The second input	$x \wedge 0$
	0	0	0
	1	0	0

7.3 The binary operator $\vee$ defines logical operation **OR**. The truth table to logical operations, $x \vee 0$ and x, is shown in Table 7.15, where for $x \vee 0$ x is a Boolean variable that is the first input and 0 is the second input. Based on Table 7.15, write a bio-molecular program to prove $x \vee 0 = x$.

Table 7.15 The truth table to logical operations, $x \vee 0$ and x, is shown	The first input (x)	The second input	$x \vee 0$	x
	0	0	0	0
	1	0	1	1

7.4 The binary operator $\wedge$ defines logical operation **AND**. The truth table to logical operations, $x \wedge 1$ and x, is shown in Table 7.16, where for $x \wedge 1$ x is a Boolean variable that is the first input and 1 is the second input. Based on Table 7.16, write a bio-molecular program to demonstrate $x \wedge 1 = x$.

Table 7.16 The truth table to logical operations, $x \wedge 1$ and x, is shown	The first input (x)	The second input	$x \wedge 1$	x
	0	1	0	0
	1	1	1	1

7.5 The binary operator $\vee$ defines logical operation **OR**, and the unary operator $'$ defines logical operation **NOT**. The truth table to a logical operation, $x \vee x'$, is shown in Table 7.17, where x is a Boolean variable that is the first input and its negation is the second input. Based on Table 7.17, write a bio-molecular program to show $x \vee x' = 1$.

Table 7.17 The truth table to a logical operation, $x \vee x'$, is shown	The first input (x)	The second input (x')	$x \vee x'$
	0	1	1
	1	0	1

7.6 The binary operator $\wedge$ defines logical operation **AND**, and the unary operator $'$ defines logical operation **NOT**. The truth table to a logical operation, $x \wedge x'$, is shown in Table 7.18, where x is a Boolean variable that is the first input and its negation is the second input. Based on Table 7.18, write a bio-molecular program to prove $x \wedge x' = 0$.

Table 7.18 The truth table to a logical operation, $x \wedge x'$, is shown

The first input (x)	The second input (x')	$x \wedge x'$
0	1	0
1	0	0

7.7 The binary operator $\wedge$ defines logical operation **AND**. The truth table to a logical operation, $x \wedge x$, is shown in Table 7.19, where x is a Boolean variable that is the first input and the second input. Based on Table 7.19, write a bio-molecular program to show $x \wedge x = x$.

Table 7.19 The truth table to a logical operation, $x \wedge x$, is shown

The first input (x)	The second input (x)	$x \wedge x$
0	0	0
1	1	1

7.8 The binary operator $\vee$ defines logical operation **OR**. The truth table to a logical operation, $x \vee x$, is shown in Table 7.20, where x is a Boolean variable that is the first input and the second input. Based on Table 7.20, write a bio-molecular program to demonstrate $x \vee x = x$.

Table 7.20 The truth table to a logical operation, $x \vee x$, is shown

The first input (x)	The second input (x)	$x \vee x$
0	0	0
1	1	1

7.9 The binary operator $\vee$ defines logical operation **OR**. The truth table to logical operations, $x \vee y$ and $y \vee x$, is shown in Table 7.21, where x and y are Boolean variables that are respectively the first input and the second input. Based on Table 7.21, write a bio-molecular program to prove $x \vee y = y \vee x$ that satisfies the commutative law.

Table 7.21 The truth table to logical operations, $x \vee y$ and $y \vee x$, is shown

The first input (x)	The second input (y)	$x \vee y$	$y \vee x$
0	0	0	0
0	1	1	1
1	0	1	1
1	1	1	1

7.10 The binary operator $\wedge$ defines logical operation **AND**. The truth table to logical operations, $x \wedge y$ and $y \wedge x$, is shown in Table 7.22, where x and y are Boolean variables that are respectively the first input and the second input. Based on Table 7.22, write a bio-molecular program to show $x \wedge y = y \wedge x$ that satisfies the commutative law.

Table 7.22 The truth table to logical operations, $x \wedge y$ and $y \wedge x$, is shown

The first input (x)	The second input (y)	$x \wedge y$	$y \wedge x$
0	0	0	0
0	1	0	0
1	0	0	0
1	1	1	1

References

J. Hopcroft, R. Motwani, J. Ullman: *Introduction to Automata Theory, Languages, and Computation.* (Addison Wesley, Lebanon,, 2006). ISBN: 81-7808-347-7

L. Null, J. Lobur, *Essentials of Computer Organization and Architecture.* (Sudbury, Jones and Bartlett Learning, 2010). ISBN: 978-1449600068

M.M. Mano, *Digital Logic and Computer Design.* (Prentice-Hall, New Jersy, 1979). ISBN: 0-13-214510-3

M.M. Mano, *Computer System Architecture.* (Prentice Hall, New Jersy, 1993). ISBN: 978-0131755635

S. Brown, Z. Vranesic, *Fundamentals of Digital Logic with Verilog Design.* (McGraw-Hill, New York, 2007). ISBN: 978-0077211646

S.G. Shiva, *Computer Organization, Design, and Architecture.* (CRC Press, Boca Raton, 2008). ISBN: 9780849304163

W. Stalling: *Computer Organization and Architecture.* (Prentice Hall, New Jersy, 2000). ISBN: 978-0132936330

Index

Printed in the United States
By Bookmasters